Parenting Through Crisis

Also by Barbara Coloroso

kids are worth it!

Parenting Through Crisis

*Helping Kids in Times of Loss,
Grief, and Change*

Barbara Coloroso

Quill
A HarperResource Book
An Imprint of HarperCollins*Publishers*

Excerpt from *Jewish Wisdom* by Rabbi Joseph Telushkin, copyright © 1994 by Joseph Telushkin, reprinted by permission of HarperCollins Publishers Inc.

Excerpt from *Bouncing Back* by Joan Rivers, copyright © 1997 by Joan Rivers, reprinted by permission of HarperCollins Publishers Inc.

"Vindication in a lonely death," an editorial published in the *Denver Rocky Mountain News* on May 27, 1998, reprinted by permission of *Denver Rocky Mountain News*.

Published in Canada as *Parenting with Wit and Wisdom in Times of Chaos and Loss*

First HarperResource Quill paperback edition published 2001

Designed by Joy O'Meara

The Library of Congress has catalogued the hardcover edition as follows:

Coloroso, Barbara.
    Parenting through crisis : helping kids in times of loss, grief, and change / Barbara Coloroso.
        p. cm
    Includes bibliographical references and index.
    ISBN 0-06-019856-7
    1. Loss (Psychology) in children. 2. Grief in children. 3. Life change events. 4. Parent and child. 5. Child rearing. I. Title.

BF723.L68 C65 2000
649'.1—dc21

00-038923

ISBN 0-06-095814-6 (pbk.)
        08 09 10 RRD 10

To my mom and dad,
Margaret Walker Reagan and Raymond Joseph Reagan,
who helped weave the rich tapestry of my heritage

For my children,
Anna, Maria, and Joseph,
that you may find warmth and love in this tapestry laced with
the threads of quiet joy and gentle peace, woven from
life's chaos, loss, love, and hope

In memory of Marty Jenco, my mentor and friend
November 27, 1934–July 19, 1996

On January 8, 1985, Father Marty Jenco was kidnapped on the streets of Beirut, Lebanon, and was held hostage until July 26, 1986. His captivity was a journey of chaos, loss, hope, and healing for himself and for all of us who call him friend.

A Servite missionary who officiated at my wedding and was the god-father of my three children, Marty was also my mentor and guide. As he was writing his book, *Bound to Forgive: The Pilgrimage to Reconciliation of a Beirut Hostage*, he would send me drafts to read. We would talk for hours about his captivity and how it had changed his life. Throughout all of the telling of the horrors and abuse, there was a gentleness and a reverence for himself and those who held him captive.

He died of pancreatic cancer and was buried ten years to the day of his release. His life, his works, his words, and his love have greatly influenced mine. This humble Servite priest taught me, a former Franciscan nun, the true meaning of the Peace Prayer of Saint Francis:

God, make me a channel of your peace.
    Where there is hatred, let me sow love;
    Where there is injury, pardon;
    Where there is doubt, faith;
    Where there is despair, hope;
    Where there is darkness, light;
    Where there is sadness, joy.

Grant that I may not so much seek to be consoled as to
console;
To be understood, as to understand;
To be loved, as to love;
For it is in giving that we receive;
In pardoning, that we are pardoned;
In dying, that we are born to eternal life.

Marty's writings are a powerful testimony to the human
spirit from a man who, bound in chains, found the strength to
be compassionate and the wisdom to forgive. His journey gives
witness to the joy and peace that come with peacemaking. I am
grateful to have shared in his journey.

Contents

Why This Book, Why Now?

Nothing is lost. All we have suffered frames a teaching for the soul.
—Bob Savino, *Song of the Floating Worlds*

The plumber has just handed you the bill for repairing the damage done by your toddler and his friend when they created a tidal wave in the bathtub.

Your eldest arrives home from college with a tattoo she assures you most people will never see.

Your fifteen-year-old wants to pierce just one more body part.

Your father-in-law phones to let you know it will be over his dead body that his Alzheimer's-stricken mother will go to a nursing home.

Your mother-in-law is dead tired after spending two hours looking for the same mother, who wandered away from the backyard for the third time this week.

Your sister's husband has left her and their three teenagers, to marry a woman five years older than his eldest daughter.

Your niece phones to ask if she and her boyfriend could stay in your guest room over the holiday because she "knows" you are so much more reasonable than her parents.

Your childhood friend turns in the manuscript for his eigh-teenth book, drives up into the mountains he loves, and hangs himself.

Your neighbors have invited you to their daughter's coming-out party and you don't think they were referring to a debu-tante ball, and it's okay and it's about time.

Your middle child's friend asks if his stepsister and half brother can ride in the car pool this week, since his father's new wife's mother is taking care of them, and she doesn't drive.

A woman down the block is having a fund-raiser to help pay for the high-dose chemotherapy with stem-cell rescue that the specialists say is the best chance her twenty-nine-year-old hus-band has to get rid of his brain tumor. The insurance company won't pay because three of its doctors say the treatment is experimental.

Across the road, the triplets born through frozen-embryo trans-plant are celebrating their tenth birthday on the same day their new brother arrives on a plane.

A child is selling his Ritalin at school, while another is selling a weapon.

A solicitation letter pleads for a donation, warning that with a click of a mouse your child can connect with a pen pal or a pedophile, and only your money can stop the madness.

Your boss called the most recent layoffs in your department "downsizing" and "rightsizing" to create a better bottom line; all you know is the only thing up is the number of people who feel down-and-out.

The four-year-old who created the tidal wave in the bathtub is now trying to show you how to override the parental-control box on the TV.

On the news, a reporter in a war zone explains that the "collateral damage" inflicted last night was simply a matter of killing the wrong persons when they were trying to kill the right enemy.

The commercial that follows demonstrates that sex can and will be used to sell almost anything; you wonder about the paradox that selling sex is illegal, buying it is not.

Ageism, racism, sexism, bigotry, religious wars, fanaticism . . . Some say it is the final time, others say it is the dawn of the new millennium; you're not sure if you are entering the dark night of your soul or beginning a hero's journey, and at this point you're too exhausted to care. You just want to sit down and have a cup of coffee, or was it tea? Which one is good for you? You can't remember.

Parenting Through Crisis is about handling tough times and tough issues with our children. Whether it be a small crisis or a major disaster, a minor malady or a life-changing event, the question "Will I die?," the loss of a pet, the death of a friend, a critical illness, a chronic disability, a divorce, an adoption, or mayhem in a high school, this book looks at how we as parents can comfort and nurture our children—and ourselves—as we navigate through the inevitable suffering, the adversity, the chaos, and the losses in our lives. It takes us further along on the journey of integration, healing, and connection begun in my earlier book, *kids are worth it!*

kids are worth it! offered practical advice for parents of toddlers through teenagers on how to utilize the very stuff of family life—chores, mealtime, sibling rivalry, toilet training, bedtime, allowances—to create a home environment in which kids can develop their own sense of inner discipline. When these routines are overshadowed by tragic or traumatic events, when children are faced with life's chaos and confusion, the lessons learned through these opportunities do not go to waste. Knowing *how* to think, not just *what* to think; feeling empowered, not controlled or manipulated; being able to distinguish between realities that must be accepted and problems that can be solved; and being able to act with compassion and integrity—these are skills that will serve your children well as they navigate the turbulent waters of adversity and sorrow.

If used in the good times, the tools of good parenting—treating kids with respect; giving them a sense of positive power in their own lives; giving them opportunities to make decisions, take responsibility for their actions, and learn from their successes and mistakes—are the same tools you can use in the rough times.

But you will need more. Suffering is a natural part of life, and we cannot eliminate it. What we *can* eliminate are the things we do to compound our own suffering unnecessarily, and the things we do to cause others to suffer unnecessarily. *Parenting Through Crisis* is a journey through the inevitable suffering that is a part of living and a journal of ways to approach with compassion and optimism some of life's most daunting situations. It is also a guide to using reconciliatory justice for handling serious mistakes, mischief, and mayhem that happen in our homes and our communities.

As a parent, I have suffered in ways that I never imagined. Others who have suffered inconsolable losses have shared their stories with me. In living through our losses and in sharing our stories, we all discovered that, although each one of us lives through a solitary grief that is our own, we are not alone in our suffering, nor are we the only ones to suffer such grave losses. We all knew this to be true in our minds before we went through the chaos. Living through it, we now know it to be true in our hearts and our souls as well.

It is our wit and our wisdom that help our children, and us, through the passages of grief. It is in our grieving that we learn a new wholeness. It is in this wholeness that we are able to embrace our sadness, knowing that it shares space with a quiet joy and a gentle peace.

As you journey through grief and suffering with your children, may you also know that quiet joy and gentle peace.

Our most important task as parents is raising children who will be decent, responsible, and caring people devoted to making this world a more compassionate place.

—Neil Kurshan, *Raising Your Child to Be a Mensch*

Part One

. . . .

ESSENTIALS
FOR THE JOURNEY

Chapter 1

Finding a Path Through Grief

Before I suffered a major catastrophe, I had no way of understanding the depth to which the soul is shaken, the exterior shattered, the interior made vulnerable and raw. Perhaps this is the way the wound works, to open us up so that we can feel and experience the depths, and having gone there, climb to heights we could never imagine. . . .

—Judy Collins,
Singing Lessons: A Memoir of Love, Loss, Hope, and Healing

Life's not fair!" Joseph sobbed as he pounded his fist on the table. He had just been told that his sister had cancer and would need a second surgery as well as radiation treatment. I could only agree. Life is not fair. I learned long ago not to ask why or why not. Those kinds of questions have no good answers and give no solace, so why bother? Why cancer? Why my daughter, who was so full of life and such a joy? Why not someone else? Why not someone older? Why not someone who wanted to die? Why not me instead of her? And why us? Only two years before, her sister was seriously ill. Now her. Would her brother be next? Such questions only drain energy from the mind and the body. We needed all the resources we could muster to get through this together.

All the colds, ear infections, cuts, scrapes, and even the broken bones of childhood seemed insignificant. All the minor hassles that had seemed so important yesterday paled by comparison. Even the deaths of the elders in the family, the automobile accidents, the surgeries seemed distant and unconnected to the suffering our family was now going through. Yet they all did somehow fit together. Getting through those traumas and losses with optimism and resolve gave us the wherewithal to approach this major one with the same optimism and resolve. We would just have to reach deeper and hang on longer.

Life hurts. We came to know a new depth to which our souls could be shaken. We met others who had been as deeply shaken. And we met others whose souls had been nearly rent apart with a suffering we had been spared. Their children died on the pediatric cancer floor that our daughter had walked away from to get on with the rest of her life. Life is good.

Life is not fair. Life hurts. Life is good. These three seemingly incompatible expressions are really three parts of the whole of living. They are threads woven through the tapestry each one of us creates as the visible expression of our being a part of humanity. To accept these three is not to abandon hope or optimism, or to deny our real grief. To accept them is to rid ourselves of the unnecessary suffering that comes from struggling against these three truths and trying to make them something they are not.

Burying grandparents, an uncle, and five friends gave our children firsthand experience that death is an inevitable part of life. In its inevitability there is suffering, pain, and grief. Our children learned that the grief has its own timetable for healing. They learned that healing does happen.

Living through the terror of the kidnapping of their godfather, Marty Jenco, gave them firsthand experience that bigotry, hatred, fear, and fanaticism have a human face and can cause just as much suffering, pain, and grief as do the inevitable losses in life. They can cause even more pain, because the suffering that was inflicted on Marty was intentional and unnecessary. After his release, our children also learned from Marty that hearts and minds can change. One of those who brutalized him in the early months of captivity came to him in the end and asked his forgiveness. Our children learned that one bound in chains can have the strength to be compassionate and have the wisdom to forgive. As he retold his story, Marty said softly, "Two men, alienated brothers, off in our own alien lands, eating the silage of bitterness and resentment, embraced. Two sons came home to their hearts, in which the spirit of peace and reconciliation lives."

This spirit of peace and reconciliation enables us to reach out to others with compassion and empathy, honoring our deep bonds and common humanity. Our deep bonds with one another give us our sense of dignity and worth in the face of adversity or a great loss. It is our compassion that demonstrates our sense of responsibility for, our commitment to, and our respect for one another. Our compassion reflects our deep passion to alleviate another's pain and suffering. Our empathy enables us to look at adversity and grief from the perspective of the one who is suffering and ask, "What are you going through? What do you need?"

Whether we are dealing with a death, an illness, an accident, a divorce, or mayhem, we will need peace of mind, optimism, and resolve to handle the chaos, the loss, and the suffering that come hand in hand with each of these. How we handle our mourning will give our children tools to handle theirs. When we offer them our compassion and empathy, we give them, from our own tapestry, strong threads of hope and resolve to grab on to and eventually weave into their own rich tapestry of life.

Life is not fair. Life hurts. Life is good.

A deep distress hath humanized my Soul.

—William Wordsworth, "Ode to Duty"

TAO of Family

A student asked Soen Nakagawa during a meditation retreat, "I am very discouraged. What should I do?" Soen replied, "Encourage others."

—from Essential Zen

Tao is the Zen Buddhist word for "way" or "path." It is not a source or an absolute. In and of itself, it yields no truth or answer. It is not *the* way or *the* path. Like an algebraic formula, Tao is both empty and useful, and like a formula, it can be used again and again in many different situations. Such is the TAO of Family. It is a path and a way.

TAO of Family is also an acronym for the three things we need when our lives are thrown into chaos: Time, Affection, and Optimism. These three form the foundation for all of the other TAOs in this book. TAO of Mourning, TAO of Illness, TAO of Divorce, and TAO of Hope—each has its own unique formula, its own way or path. But they all start with time, affection, and optimism.

Time

I expect to pass this way but once. Therefore if there be any kindness I can show or any good that I can do let me do it now.

—William Penn

Parenting is not an efficient vocation. It takes time. And when we are consumed with grief, it is often difficult to find time for anything except our own grieving. We hope our kids will see our grief and understand. Understand, maybe; accept, probably not. We need to find time for our kids, even if it is time to share in the grieving, lest they become the hidden mourners.

When we are dealing with our own trauma during a divorce, we might be inclined to assume that our resilient kids can hang in there by themselves until we put our own lives back together.

They can't. When one child is seriously ill and taking up most of our time and energy, we might hope the other children will patiently wait for this latest crisis to pass before they can expect to get their needs met. They probably won't.

Kids need some of our time every day in the good times. They need that time even more in the rough times. It's not necessarily a lot of time—just some, to know that they are listened to, cared for, and are very important to us. The time doesn't have to be spent doing something planned or special. It can be "hanging time," just being around for the kids and not hiding out in our darkened bedroom. "Hanging time" can be driving the car pool, getting everyone out of the house and doing something together that is routine and normal—a break from grief.

Spending time with our kids can help them handle their own mourning. We also need to *give* them time to get through their grieving. There is no way to rush grief, condense it, or eliminate it. If we don't give kids the time now, they will need to take the time later. Grief doesn't just go away. It must be gone through. You might be happy to be remarrying and excited to get on with your new life, but you can't expect your seven-year-old to share immediately in your joy. She might need to grieve the now-shattered dream of Mom and Dad getting back together, grieve the loss of the single-parent home, grieve the space she has to give to her stepsister.

The siblings of a critically ill child need a break—a break *from* their sick brother and a break *with us* to gain back some semblance of normality, to climb mountains and not feel guilty because their brother can't. The ill child needs a break *from* all of us to spend time alone or *with others* who are going through the same treatment he is enduring.

We need to take time to be silent, to think, reflect, and just be. Our children also need that time. And we need some quiet time together, to be still in our grief: willing to be present and not act. In stillness we can be more aware of a bigger picture. Sometimes possibilities that didn't present themselves in the midst of a crisis come forward during the still moments. And sometimes possibilities that didn't exist at the moment of the crisis come together to create an even better resolution than was even possible in the first hours, days, or weeks of a loss. In our stillness we can be open to those possibilities.

It all takes time.

Affection

Receive every human being with a cheerful countenance. Rise to the occasion when no one else will.

—from the Talmud

Our children need a smile, a hug, and humor every day. In times of grief, these three are often cast aside as a gray, cold heaviness descends upon the house. But it is these three that can help all of us get through our mourning. A smile, even one we had to work hard to create, lifts our spirits. Hugs let us know we are in this together. A hearty laugh is contagious and can provide a respite from our grief. With these three simple gestures we give our children all three parts of the TAO of Family: our time, affection, a sense of optimism—and we do it with little thought or effort. Which is a good thing, because thought and effort are usually in short supply when we are grieving.

Along with a smile, a hug, and humor, we need to give our children unconditional love. Unconditional love is just that: to love without conditions. It is to hang in there with our children through the good times and the rough times. It's not "If you're well behaved, I'll love you; if you're not, I won't."

If we become grandparents sooner then we would ever have hoped or wanted to be, and our child, who is just learning to change the oil in the car, is also learning to change his own baby's diapers, we need to be there. When our teen calls us from the local jail to let us know why she's not making it home before curfew, we need to be there. When our young adult is diagnosed with a mental illness and her erratic behavior is driving us crazy, we need to be there.

Our likes and dislikes can be highly conditional. We don't have to like our children's hairdos, strange-looking shoes, earring in the nose and navel, obnoxious behavior, or chosen field of study. Our love, on the other hand, has to be something they can count on. Our being there is not to condone, excuse, or support our children's behavior. It is to encourage, provide feedback, provide a supportive presence to them, and simply to love them.

> . . . that's a parent's job: to love without qualifications, to embrace without any conditions.
>
> **—William Ayers, *A Kind and Just Parent***

Optimism

I have learned silence from the talkative, tolerance from the intolerant, and kindness from the unkind. I shall not be ungrateful to those teachers.

—Kahlil Gibran, *The Prophet*

It is a grateful attitude, a willingness to view even adversity and adversaries from a fresh perspective, that is the hallmark of genuine optimism. It is the ability to go through a long night of grief, get up in the morning, make breakfast for our children, and affirm to them that all of us can make it through this. Such resilience will not come to us in times of great chaos and grieving if we haven't made it part of our way of approaching everyday ups and downs. How do we respond when a colleague is late, when a child breaks a glass, when we are tired, worn out, and frustrated and our child announces he needs to read three books out loud to us tonight? Optimism doesn't deny anger, frustration, sadness, or intense sorrow. It is willing to give each one its due, but only its due. We cannot always control what happens to us, but we can control how we *respond* to it and how we *use* it.

It is not easy to respond with optimism when faced with a hostile letter from our children's other parent, or a call from the local police, or a diagnosis that shatters our hopes. To accept realities for what they are, look at ways to use those realities for good, and get busy solving the problems created by those realities helps us reaffirm our optimism.

Every time we reaffirm our optimism, we give our children a good way to approach their own adversity. They can take an active part in determining what they will do with what life has handed them. They will be less likely to be passive recipients of whatever comes their way. They know how to view change, be it welcome or unbidden, as a challenge and an opportunity to grow. It is this perspective on life that the philosopher Reinhold Niebuhr spoke of when he prayed, "God, give us grace to accept with serenity the things that cannot be changed, courage to change the things which should be changed, and the wisdom to distinguish the one from the other." An optimist knows the difference and works all three to the greater good.

The Triangle of Influence

The eye, the mind, the soul, each has its own perspective.

—Maori proverb

In times of chaos, if we are going to respond to our own suffering and our children's grief in an active, self-aware, compassionate way, we will need to use our minds, our hearts, and our intuition together. To rely exclusively on any one of these to get through the suffering is to narrow our perspective, limit our options, and hamper our grieving.

If we use only the mind, we will end up trying to define the suffering with logic. We will make a valiant attempt to give reasonable explanations for the pain. We will try to find some purpose for the grief and a plan for ending it. We will look to stages and timetables for definitive benchmarks for our suffering. They don't exist. And if we give our children the idea that they do, we will be doing them a disservice.

Suffering is anything but logical or reasonable. Seeing suffering as a problem that can be fixed with a plan might cause us to act hastily, just to be *doing* something. In reality, what might be needed is simply being present for our children and ourselves and being open to the pain and suffering without trying to "fix" either.

We can consciously busy ourselves to repress our pain and distract ourselves from our suffering. We can waste time asking unanswerable questions. Establishing a goal will frame our suffering as something that can be completed, gotten through, and finished. Grief doesn't work that way.

Acting alone, logic might also invite us to play the blame game: If we are suffering, someone or something has done this to us. We are victims of a malevolent person, a malevolent virus, a malevolent universe. Rather than grieving, we spend our energy trying to get back at those someones or somethings to make them pay for the pain they have caused us. We tally up the inequities, enumerate the hurts. Our suffering continues. We can seethe with hatred and curse others relentlessly for causing us pain, knowing deep down that if we let go of our hatred, we will have to face that hurt, mourn its pain, and heal. And our children, trying to follow the example we set, will be locked in their own grief as well.

The heart alone does no better job. The philosopher Blaise Pascal observed, "The heart has reasons which reason knows nothing of." These "heart reasons" are not logical. They can't be rationalized or scrutinized. Separated from any logic, they might very well sound unreasonable.

If we look only to the heart for a way out of our suffering, the overload of emotions can cause us to act impulsively. We can overwhelm ourselves and others with our grieving. We call it a mental breakdown when in reality it is an emotional meltdown. We slip into a deep depression and refuse to acknowledge that life is good. We waste our time worrying about past transgressions and future calamities.

We can try to numb our pain with alcohol or drugs.

We can get so riled up with anger and vengeance that we create greater suffering for ourselves and others. We call it a "crime of passion."

We can also use our "heart reasons" to excuse behavior in an effort to prevent us from having to experience the pain and suffering that comes with owning up to what we have done. We might rush in to rescue our children from the consequences of their own choices, decisions, and mistakes for the same reason. In these ways, we fail to show our children how to take responsibility for their own actions.

When either mind or heart works independently of the other, the denial, repression, hatred, blame, and worry created rob us of peace of mind, our sense of optimism, and the resolve we need to face our suffering and heal our pain—and help our children do the same.

It is our intuition that can bridge the two seemingly disparate perspectives. Intuition is often called "the voice of the soul." Being able to acknowledge, trust, and act on our intuition is particularly useful when we are faced with complex difficulties, major chaos, and profound loss. It can point a way out of an impasse that thinking and emotions have created by doing battle with one another. It often provides options that aren't immediately obvious to our mind or our heart. When we are in touch with our intuitive self, we have more choices. When we connect our head and our heart to our intuition, we are no longer just logical or just emotional. We no longer merely *react*. The head-heart-intuition connection forms a powerful triangle of influence. We become *mindful* with a *wise heart*.

Our intuition helps us to know when to reach out and when to refrain from reaching out, when to speak and when to be silent, when to hold on and when to let go. In a small or large crisis, we are able to *respond* with a generous spirit, wisdom, discernment, empathy, abundant kindness, mercy, and compassion. It is all of these that we will want to rely on continually as we journey with our children through the inevitable suffering, the adversity, the chaos, and the losses in our lives.

With compassion, we see benevolently our human condition and the condition of our fellow beings. We drop prejudice. We withhold judgment.

—Christine Baldwin,
Life's Companion: Journal Writing as a Spiritual Quest

Part Two

· · · · · · · · · · · · · · · · · ·

THE BIG ONES

Chapter 2

• • • • • • •

Death: Helping Kids Mourn

Grief is a tidal wave that overtakes you, smashes down upon you with unimaginable force, sweeps you up into its darkness, where you tumble and crash against unidentifiable surfaces, only to be thrown out on an unknown beach.

—Stephanie Ericson, *Companion Through the Darkness*

A grandparent dies after a long illness. A parent is killed in a horrific accident. A brother or sister is rushed to a hospital and never comes home again. A classmate commits suicide. Confronting the reality of death honestly and directly with children is difficult at best. In a death-denying, fix-it-fast, cure-it-now society, with so many rituals and customs of our ancestors abandoned or never experienced, the task is even more painful and necessary. Our own feelings, belief systems, faith traditions, questions about morality and the meaning of suffering—as well as our understanding of the abilities of children to handle loss and grief—can help or hinder us in helping our children mourn. We can try to hide the loss from them, try to shield them from the anguish, convince ourselves they are too young to understand—they will still grieve, but without the comfort, support, knowledge, and tools they need.

As you deal with your own grieving, shock, disbelief, and anguish, you won't be inclined to sit down and read a book called *101 Ways to Help Your Children Cope*. But there are things you can do in advance of a

death and during the passages of grief that can help both you
and your children journey the uncharted waters of your loss.

*We act like life was certain and death uncertain. Life is uncertain and death
certain.*

—the Reverend Jesse Jackson in a eulogy for Ennis Cosby

The Circle of Life

*In the first moment when we come away from the presence of death, every
other relation to the living is merged to our feeling, in the great relation of a
common nature and common destiny.*

—George Eliot, *The Lifted Veil*

Before they face the death of a family member or friend, it helps
if children first learn about death from everyday events such as
the changing of the seasons, a dead bird in the yard, the death of
a family pet. Observing life cycles in everyday living, and talking about them matter-of-factly, can be one kind of preparation
for the inevitable deaths of loved ones.

Given the opportunity, children will ask questions such as
"What is death?" "Does everything have to die?" "Why did the
dog die?" "Will you die?" "Will I die?" "When will Grandpa
die?" "What happens after death?" "Where will we go after we
die?" It is easier to explain the basic attributes of death and
answer the inevitable questions calmly and forthrightly when
we are ceremoniously burying the dead robin than when we are
grieving at Grandpa's grave.

Don't tell your five-year-old that the dead bird on the win-
dowsill is just taking a nap or that the dead goldfish is practicing
the back float in the fishbowl. Don't say that Mufasa didn't really
get killed in *The Lion King* or that he will return in *Lion King II*.

Dead is dead. It does not lend itself to adjectives—except in movies such as *The Princess Bride,* when the irreverent Miracle Max (Billy Crystal) declares the hero "only mostly dead."

A death that touches any member of a family has a significant impact on all members of the family. Even toddlers can and do mourn a death. They may not have the language, but they do have the feelings and intuitive sense of loss. In his preface to *What Maisie Knew,* Henry James reminds us that "small children have many more perceptions than they have terms to translate them; their vision is at any moment much richer, their apprehension even constantly stronger than . . . their at all producible vocabulary." Even in infancy, children's feelings and their intuitive sense can be complex and expressive when their language is not. All children who experience the death of a family member feel helpless and lost. At all ages and stages of development, children have ways of coping with loss. Even when they are too young to understand the concept of death or speak what they are feeling, they are able to grieve.

Children as young as four can begin to understand that living things have, as Bryan Mellonie and Robert Ingpen wrote in their book *Lifetimes: The Beautiful Way to Explain Death to Children,* "beginnings, and endings, and there is living in between." Children as young as five can begin to understand what Elliot Kranzler describes as the four attributes of death (text in parentheses added by author):

1. It has a specific cause (nobody just "drops dead").
2. It involves the cessation of body functions (the body can't move, can't feel, can't breathe, can't grow—it's not just sleeping).
3. It is irreversible (it can't be undone, there are no "overs").
4. It is universal (it happens to all living things).

As your children are exploring the concept of death, you can reflect on how you as a child learned about death. How were the questions you asked answered? Were they even asked? Did people close to you die? Did you take part in any of the rituals, burials, services, memorials, or commemorations for them? If not, why not? Were secrets kept about any death in your family—a suicide hidden, a disease not mentioned? Did any pets die? How was the death handled? Are there any deaths you feel you have not grieved?

In coming to terms with the newly dead, I seem to have agitated the spirits of the long dead. They were stirring uneasily in their graves, demanding to be mourned as I had not mourned them when they were buried. I was plunged into retroactive grief for my father, and could no longer deny, though I still tried, the loss I'd suffered at the death of my mother. . . . Was it possible . . . that one could mourn over losses that had occurred more than half a century earlier?

—Eileen Simpson, *Orphans: Real and Imaginary*

Passages Through Grief

The first piercing grief eventually becomes a kind of ever-present sorrow that doesn't seem to want to go away ever, but it does; or rather, it grows into something else, something you know you can live with, although at the same time you know you will never forget.

—Chuck Norris,
discussing the death of his brother Weiland in Vietnam,
in *The Secret Power from Within: Zen Solutions to Real Problems*

There is no destination, no arrival, no ending place in the journey of grief. There is no road map to follow, no formula, and no way to hurry the journey or bypass the pain. There are passages to live through, not stages that we move past in a lockstep, hierarchical order. To force ourselves or our children into a linear grieving "process," evaluating where we are on the ladder of grief, is a vain attempt to control and manipulate a "journey of the heart." This journey cannot be controlled; it can only be lived through by each one of us in our own time and in our own way.

In *Living Our Dying,* Joseph Sharp, intern chaplain to patients diagnosed with terminal illnesses, describes the journey as a circle with passages along the edge. "There is no actual upward progression along the circle; there is only a coming around. . . . Our prime obligation is to be forthright and honest about opening into what is directly before us. The circle directs us back to the present moment." In a sense, it directs us back to ourselves in coming to know our own living and dying more fully. Dr. Elisabeth Kübler-Ross suggests that we have two choices, "to live in grief, remorse, and guilt . . . or to face those feelings, work them through, and emerge with an acceptance of death and a commitment to living."

As we journey through our many losses in life, there are three passages that we experience over and over again:

1. The piercing grief of good-bye
2. Intense sorrow
3. Sadness that shares space with a quiet joy and a gentle peace

When someone dies, we and our children, all in our own time and in our own way, need to go through these passages if we are to choose "an acceptance of death and a commitment to living." There is no one way, no right way, no only way, no singular journey. We can share our path with others, but in the end it is up to each one of us (our children included) to create our own path.

It is important to remember that these passages are fluid and overlapping. They don't always take place in a set order. In the book *Losing Uncle Tim,* by MaryKate Jordan, a young boy learns that his favorite uncle is dying of AIDS. The boy moves from an initial sadness, still filled with hope and disbelief, to an intense sorrow as his uncle Tim's weakened body attests to the fact that he is dying, through the piercing grief of his death and funeral, on to a renewed sorrow, and then finally to the sadness that shares space with a quiet joy and a gentle peace as the boy sits in front of his window. "The quilt is right here on my bed. The duck sits on my desk, next to the checkers set. Maybe when I grow up . . . I'll own a store. Or I might do something else. I don't know what yet. But I'll do something I love. Just like Uncle Tim."

All three passages can be present in the same day. Feeling numb from the shock of the sudden and unexpected death of her elderly father, Denise was rocking in his favorite chair as others around her were reminiscing about the past, trying to draw her into their stories. Denise's son, himself reeling from the death of his grandfather, asked Denise to rock her grandson, William. Her piercing grief flowed into a sadness tempered with joy as William cuddled next to her and asked her to sing his favorite lullaby, one Denise had learned at her own father's knee.

All three passages can, one by one, barge into our lives unbidden and without notice. Just when he thought he had moved through the grief and the sorrow, relishing the joy and peace of getting on with life, Dan heard a song that brought back that same grief, the numbness, the hole that was created five years before by the accidental death of his teenage son, Jimmy. The grief is no less intense, the sorrow as visceral as it was when that song was played by his son's friends at the funeral.

The path we each will crawl, walk, run, stumble through, sit down on, and at times try to run from will be our own. Others can tell us of their own journey, where the potholes were, the ruts, and the resting places, what they carried and what they discarded, wished they had brought along, or found in hidden stashes along the way. In the end it is up to each one of us to travel our own path, naming our loss, honoring our grief, confronting our pain, and telling our story.

1. The piercing grief of good-bye

Grief is not a disorder, a disease or a sign of weakness. It is an emotional, physical, and spiritual necessity, the price you pay for love. The only cure for grief is to grieve.

—Earl A. Grollman, *Straight Talk About Death for Teenagers*

This first passage is marked by numbness and shock. The body mercifully provides us with these two to help us slowly face the impact of our loss. We might feel as if we are walking in a dream state, appear to others as being stoical or robotlike. We are often in denial, hoping to wake up from a nightmare, searching the papers or listening to the news, fighting, against all odds, to learn that it isn't true. The first words out of our mouth—"Oh, no"—are a frantic attempt to change what is.

We may need someone else to verify to us that the death is real. In her book *A Mother's Memoir*, Gloria Vanderbilt writes about the day she watched in horror as her twenty-two-year-old son, Carter Cooper, took his own life by leaping off a ledge fourteen stories high. Even though she actually witnessed him letting go of the ledge, she urged her housekeeper to call the police. "Yes, yes, I urged, we must call them, do that, they are the authorities—*they* would be the ones to tell us *it hadn't happened*. I too started calling—Dr. Young, certain that he too, an even higher authority than the police, would tell me *it hadn't happened* . . . Suddenly there were sirens, and the room was full of people. Dr. Tuchman, our family physician, was standing in front of me. *He* would be the final one to know . . . *Tell me it's not true?* I looked into his eyes and knew then that *it was true*."

Nine-year-old Sam's father was killed in a mine explosion. Sam insisted on watching over and over again the news clips of the explosion and futile rescue attempts. Five days after the explosion, he walked up to the TV set, turned off the news as it recycled clips of the disaster, walked back to the couch, wrapped himself in his father's favorite sweater, said to no one in particular, "My father's dead now," and started to weep.

During the first few days following the death of someone close to your children, they might feel isolated, confused, and scared, their routines disturbed, their lives irrevocably altered. Talk to them about what will happen the next hour, today, tomorrow, for the rest of the week.

Children do as we do: feel too much or not at all, cry a lot or not at all, play the "what if?" mind games, go over and over the details of the death trying to change the ending, wanting to eat, not wanting to eat, wanting to be alone, not wanting to be alone. Such is the nature of shock. Logic does not make sense; there is no order to this chaos.

Share your feelings with your children and let them express their own. Your words can help them give expression to their own emotions. Know that, along with shock, all of you might feel anger, rage, and panic. These emotions are real and need an outlet for expression that doesn't harm anyone. Give your children time just to be, to cry, to think, to talk, to question, to help, and yes, to complain. Life is not fair.

From the time of a death until the funeral, it is the grief that swamps all of us. The whole world seems unpredictable, unstable, and unmerciful. True mourning usually begins at the wake, funeral, or memorial service. The order and customs inherent in these ceremonies can help children find some sense of calm in all of the chaos. Don't force your children to attend these ceremonies, but do encourage them to go. It is important that they know in advance what to expect. Don't know what to tell them? Just go back to the six fundamental questions: Who?

What? When? Where? Why? and How?—filling in the blanks as they relate to your own cultural or faith traditions.

Even young children can take part in the rituals, drawing pictures or writing letters to the person who has died, picking out a flower to place on the grave. Older children can help plan and take part in the ceremonies and celebrations. Ask for their ideas, and respect the thoughts and feelings that make up those opinions: "When Johnny was alive, he always liked to wear that baseball hat. I don't want his dead body buried with it. I think he would rather my alive body wear it." Give them simple errands to run or tasks to do that are truly helpful, not an excuse to get them out of the way. Even toddlers can put a picture of Grandpa on the table. (When my father-in-law, Dominic, died, our children helped put pictures of their grandpa on the dessert table. As his friends and family helped themselves to cake, remembering Dominic's love of sweets, many would chuckle at Dominic's permanent place next to the dessert.) Older kids can help answer the phone, put the trays of food in the refrigerator, help find a suitable tie for their uncle, who had said often it would be over his dead body that he would wear a tie, and now wants to wear one to his father's funeral.

It is important to include in the ceremonies, rituals, and storytelling others who have been touched by the death you are grieving. Grandpa might have been closer to his veteran buddies than to most of his distant blood relatives. A teenager's girlfriend needs as much, if not more, comfort, support, and opportunity to grieve openly as the cousins do. In the movie *One True Thing,* Meryl Streep's character, Kate Gulden, leaned on her friends, and they on her, during their times of crisis and celebrated in their times of joy in a way none of their family members could understand or fully appreciate. At her death, those same friends added their own special touches to the wake and funeral that honored who she was to them.

Human touch at this time can be healing touch—a back rub, foot rub, head massage, a hug, a kiss, holding on to and holding up one another. Massage is medicine for our heads, our hearts, and our bodies. It can help reduce agitation and stress. It is known to increase the endorphins in our brains, natural pain relievers that we so desperately need in times of chaos and loss. Alcohol and drugs only mask the pain that must be eventually faced; our endorphins soothe the pain. Instead of reaching for a scotch, reach out and touch one another. Touch helps both the giver and the receiver; as you are soothing your upset child, you soothe yourself as well, creating a peaceful space for both of you.

During this time, it is important to give shelter to one another as each of us individually, but also as part of a family and a community, slowly continues the circle journey from the depths of piercing grief to intense sorrow, a passage that has its own elixirs and its own dragons to slay.

2. Intense sorrow as we reorganize our life

I do not believe that sheer suffering teaches. . . . To suffering must be added mourning, understanding, patience, love, openness and the willingness to remain vulnerable.

—Anne Morrow Lindbergh, *Hour of Gold, Hour of Lead*

The mind is no longer on hold, the reality of the death is seeping into the very marrow of the bones, the numbness is wearing off, a dull, constant pain taking its place. In piercing grief you faced something awful, but thankfully you could not begin to comprehend the full extent or impact of the devastation.

The nightmare of wakes, funerals, cemeteries, graves, caskets, crematoriums, and headstones now gives way to the logistics of everyday routines, routines that are the same and yet irrevocably changed, now colored stone-cold gray. The sorrow envelops your mornings, evenings, and nights, allowing no respite. The impact of the loss hits you like a ton of bricks at every turn. The empty place at the dinner table. The socks in the laundry. The birthday card unmailed. The phone call that doesn't come. The game not played. The bedtime story not read. Even happy times bring you sorrow.

The "Oh, no" of the first passage gives way to the nagging "Why?" The attempt to understand is fruitless, because no amount of understanding will dull the pain. When children ask you the "Why?" you have asked yourself a thousand times, know they are not just looking for an answer, because no answer will ever be good enough. They are looking for ways to get rid of the pain, make the sorrow go away, lift the grayness in their lives. Death is illogical and often senseless. You can't reason the pain away.

Nor can you rush through this passage or deny it its due. The sorrow needs to be expressed. Sometimes it is spoken through art or poetry or dance or running wildly through the woods. Sometimes it is just there in the silence at the breakfast table.

Often in this second passage, the "Oh, no" gives way to a feeling of anger. Anger is not in and of itself good or bad. The anger at a brother for driving too fast, a grandmother for smoking, a sister for getting cancer, a God who would let this happen is a real and palpable anger. This anger needs to be understood and expressed in a healthy way. Your children need to know that they can speak this anger without fear of being castigated or reprimanded. When they speak it to you, you can help them explore its roots. Where did it come from? *(From inside myself)*. It is masking another feeling? *(I am hurt, or frustrated, or disappointed, or afraid)*. Why be angry at all? *(Because I care. If I didn't care, I wouldn't be angry. I can't be angry about something I don't care about, with someone I don't care about.)*

After exploring the roots of their anger, your children can judge whether the anger is an appropriate response to the situation—does it give them energy to take a strong stand on an issue or make important changes springing from a feeling of compassion or a sense of concern? If so, the anger can serve them well. Fighting for stiffer drunk-driving laws, getting a curve in the road eliminated, taking better care of their own bodies—these can be constructive ways to use the energy of the anger and transform that anger into something constructive.

However, your children might want to express their anger in ways that are violent or aggressive. This will only increase their pain and suffering. No amount of anger will protect them from the hurt or injury or harm that has already been done; it will only add to the anguish. They might want to repress their anger. Feeling overwhelmed, helpless, and incapacitated, they may stuff their anger deep in their minds and bodies. Such anger unexpressed can easily turn to bitterness and pessimism.

Often the most constructive thing that can be done with the anger your children are feeling is to go back over the three questions—Where did it come from? Is it masking another feeling? Why be angry at all?—and be still with the anger, realizing that it is probably masking deeper feelings of caring and loss. Mindfully analyzing the thoughts that trigger the feeling of anger, and with a wise heart looking at them from different perspectives, can help dissipate the angry feelings and allow compassion to come through. This compassion does not eliminate suffering; it only makes it more bearable. Choosing not to act on the anger yet not deny the anger can sometimes enable your children to avoid becoming overwhelmed by the loss and to continue through their passages of grief.

It is hard to understand how people around you can possibly feel pleasure, laugh, and go about their everyday lives. Your life has been torn apart, and the rest of the world doesn't seem to notice. The people who were there for you during the early days of mourning don't even bring up the name of the dead person anymore. They seem to have picked up their pieces and appear to be impatient that you haven't even begun to pick up yours. You and your children need to speak of the loss, tell your stories over and over again. Speak again and again the name of the one who has died to make that person present now in a new way. Be as patient with your children's stories as you wish your friends and relatives would be with yours.

Nina's son, Sean, was killed by a drunk driver. For weeks after the funeral, Nina set a place at the table for her only son. Her husband and daughter were concerned that she was stuck in grief. A month to the day after the burial, Sean's best friend came to visit and joined the family for dinner. As the boy sat at Sean's place at the table, Nina's husband and daughter held their breath, waiting for Nina's response. Smiling, she handed Sean's friend a plate and asked him to tell her about the good times the two boys had shared. Nina continued to set an extra place at the table, and it was soon known in town that it was for any of Sean's many friends to come and share their sorrow in losing their friend and their joy at having had him as a friend. Nina had provided for her family and Sean's friends the opportunity to speak their stories and speak his name.

Adults and children need others to listen to them, talk to them, hold them, and care for them. Five-year-old Kenny could barely drag himself out of bed a few weeks after his older sister died. His grandmother, having herself lost a child, was there to help him get dressed, fix him breakfast, and read him his bedtime story. She was also there the first month to take over for Kenny's mother, who was so consumed with her own grief that she would forget to eat if someone did not put food in front of her.

During this second passage, the first anniversaries will need to be gotten through—holidays, the anniversary of the death, and also special days that have meaning to you. Often the anticipation of these days is more painful than the days themselves. If the death was from a suicide or an accident or violence, there might be two anniversaries: the day of the death and the day the body was found or recovered. Both dates are forever seared in your memory.

Make plans to do something that will help you and your children get through each one of these days in a meaningful way. The first year you might all want to do what you would have done had the person who died been alive. Or instead, you might want to forgo the family tradition and do something totally different; maybe you'll decide to return to your traditions the following year, maybe not. To pretend the day is just another day is to deny the loss and the grief. Grief will sneak up on you anyway. So expect it and plan for it.

If your child is grieving the death of a friend, help him find ways to honor special days in a manner that is meaningful and helpful to him. Offer to be there with him if he would like, but respect his privacy if he would rather do it alone.

After the accidental death of her friend, Maria put flowers on his grave every month, without fail, on the numbered day of his death. During the second year and the years after, she made a special effort to place flowers on his grave on his birthday and on the anniversary of his death. The only participation she wanted from me was the tacit approval of her taking fresh flowers from the vase on the kitchen table and not saying a thing about their being gone.

It is during this second passage that we slowly begin to feel a desire to be done with this disorganization, this disorientation, the sorrow and pain. One way to move through it is to reach out to others who are suffering. One father found that delivering meals to AIDS patients was a way to deal with his own grief. It gave him a reason to get up and get dressed and get involved in life again. If the death was the result of an illness, another way to move through this passage is to become involved in fund-raising activities to help find a cure. Knowing how helpless you felt and not wishing another parent that same feeling, you can begin to find new meaning in your own life. Your older children might also want to actively take up a cause that now has personal meaning to them. Anna climbed Mount Rainier for the American Lung Association, soliciting money even from her grandfather who was suffering from emphysema. No one could get Grandpa to quit smoking, but Anna could get him to put money into research to find a cure for emphysema.

It takes courage to get through this passage, to not deny it, inhibit it, or rush it. You will find if you confront the pain honestly and directly and you are open to its lessons, you will increasingly feel the desire to let go of the intensity of your grief and get on with your own future.

I still miss those I loved who are no longer with me but I find I am grateful for having loved them. The gratitude has finally conquered the loss.

—Rita Mae Brown, *Starting from Scratch*

3. Sadness that shares space with a quiet joy and a gentle peace as we recommit to life ourselves, tempered by the loss and wiser

By "coming to terms with life" I mean: the reality of death has become a definite part of my life; my life has, so to speak, been extended by death, by my looking death in the eye and accepting it, by accepting destruction as part of life and no longer wasting my energies on fear of death or the refusal to acknowledge its inevitability.

—Ettie Hillesum, *An Interrupted Life: The Diaries of Ettie Hillesum*

Not feeling bad for feeling good is a sign you are moving into the third passage of grief. Tired of being tired, ready to get on with your life, and no longer preoccupied with despair, you laugh more and are able to concentrate better. Perhaps the best sign that you are ready to recommit to life is that your memories of the person who has died bring you more warmth than pain. Your relationship with the dead person has changed, but you now know it didn't die. That relationship has "carved a holy place deep in your soul," a place you can go to in the quiet solitude that eluded you in the second passage.

Almost two years after Marty Jenco died, a mutual friend sent a quotation from St. Bernard of Clairvaux: "I can never lose one whom I have loved unto the end; one to whom my soul cleaves so firmly that it can never be separated does not go away but only goes before." "Only goes before" are hopeful words that can be truly understood only by those who have mourned. They are the words that speak of the paradox of pain and grace.

No longer plagued with the "Why?" that has no decent answer, you move on to "What will *I* do now?" You've said your good-byes, you've restored yourself, and now you are ready to reinvest in your own life. The sadness is there, but it shares space with the quiet joy and the gentle peace.

Dorothy Foltz Gray, a twin whose sister was murdered, wrote in her story, *Now I Am One:*

> So I find myself unexpectedly lucky. I am often lonely, but I am not unhappy. Sometimes I still believe I live for both of us, but it is not an idea I allow to grow too vivid. It is easier to imagine, as Deane and I always did, that my part and hers continue to intertwine . . . I am alive, but like Deane, I do not live the life I started.
>
> My friends ask if her loss is with me every day. It is a hard question to answer. The day she was shot down marks a line in the center of my life. It changes the way I see those two redheaded girls who rode wooden stick horses and slept in blanket forts. It changes the way I'd pictured us as old ladies, toting cookies in one hand and grandchildren in the other. But I get to live, and life does grow back—or at least forward.

Your children might move into this third passage before you or after you. They need your support and reassurance that it is good to get on with their lives. Let them know that getting on does not mean forgetting, trivializing, or getting over the death of their loved one. It means always remembering, honoring the relationship that is there, and knowing that one does not get over a death but gets on with life.

Mourn not too long that he is gone, but rejoice forever that he was.

—ancient proverb

Humor
And if I laugh at any mortal thing,/'Tis that I may not weep.

—Lord Byron, *Don Juan*

Humor is potent medicine for the heart, the body, and the soul. It releases tension and provides us with energy to deal with feelings that could easily overwhelm us. It is life-affirming. It is a tribute to the one who has died, a celebration of the joy once shared. Children laughing and playing in the yard, getting on with life after a funeral, can lift our spirits. A funny movie can jar loose our sadness and let us feel a moment of joy in the midst of great sorrow. A humorous anecdote in a eulogy can prove to be a great antidote for seemingly unending pain. Upbeat music can get us to laugh through our tears.

At the service for Bob Collins, our son Joseph's art teacher and mentor, who was killed in a freak automobile accident, grieving family, students, and friends sat sobbing and subdued as "Amazing Grace" was played. The minister solemnly welcomed everyone and asked us to stand to sing the one hymn he knew Bob would want for the beginning of his memorial service. The organist led us all in a wonderfully boisterous rendition of "Take Me Out to the Ball Game." Tears and laughter flowed together.

We are all grateful for people whose humor in the time of loss can give us a jump start at reconnecting with life. However, the wit that helps us through a funeral—that same wit that helps us deal with our two-year-old's artwork on the newly wallpapered wall, the sandwich in the VCR, and the strangely tinted laundry—might take a turn toward the humorously absurd when we are face-to-face with death and its aftermath. Those around us who haven't yet found their own funny bone—or who wish us to handle grief in a solemn, prescribed, or at least dignified manner—will be stunned, shocked, and aghast at our gallows humor.

In her book *Bouncing Back,* Joan Rivers, talking about her husband's suicide, said, "my feelings were so overwhelming that I was able to deal with them only in the way that I had always dealt with pain: by laughing through my tears."

One night, after the week of shiva had ended, Melissa and I walked into a Los Angeles restaurant. Everyone was staring at us, waiting for us to order hemlock cocktails or stab ourselves with our salad forks. But we held our heads high; and when I opened the menu, I said, "If Daddy saw these prices, he'd kill himself all over again!"

For the first time since the suicide, Melissa laughed. Anger left her eyes, and I saw the old sparkle there. I also saw every head in the restaurant turn toward us disapprovingly.

Her husband dead a week and they're *laughing*!

These people don't want to see us heal, I thought.

They want to judge us on their own terms. They're angry that we're breaking the pattern of mourning.

Just as we adults use gallows humor to help handle our grief, our children may be inclined to do the same. As Jacob and Sarah tried to explain to their three children that Jacob was HIV-positive, the older daughter left the room screaming, the younger daughter wept, and the nineteen-year-old-son threw up his arms and sighed, "Oh, great, you always said you would try to be a positive influence on our lives. This is one kind of positive I hadn't counted on." Both dad and son managed a laugh, then hugged one another and sobbed. The attempt at humor had helped both of them break a barrier of silence that had existed for the last few months while they both skirted the reality—dad fearful to tell and son knowing something was wrong.

I was always one laugh away from crying.

—Red Skelton, comedian,
whose son died of leukemia at the age of seven

"Grandma Was on the Roof..."

A child can live through anything, so long as he or she is told the truth and is allowed to share with loved ones the natural feelings people have when they are suffering.

—Eda LeShan

Even if you have talked with your children about death and dying, when faced with the death of a loved one and confronted with your own grief, there is no easy way of getting through breaking the news. A recently widowed friend, whose father died when she was eight years old, told me she knew what *not* to say or do with her own children when their father, her husband, died of a brain tumor. She wouldn't hide the unpleasant truths; wouldn't wrap those truths in sentimental, philosophical, or religious clichés; wouldn't whisper to other adults when the kids were present; wouldn't tell her kids not to cry; wouldn't keep them from the funeral; wouldn't pack up all their dad's belongings and take down all his photographs. These things she knew she *would not* do. It was what she *would* say and do that was the problem.

Use simple, honest words: Daddy died. Your sister was killed in a plane crash. Grandpa died last night. Your aunt killed herself. Honesty doesn't have to be cold and harsh and unfeeling. Your tone of voice, what you say, and how you say it can speak warmth, caring, and sadness.

Often parents feel they have to soften the blow by beating around the bush before they get to the fact of the death. I am reminded of the story of the cat on the roof. Upon returning from a holiday, a man is greeted by his younger brother, who hugs him and cries, "Your cat died while you were away." The elder brother scolds the younger one: "You could have at least softened the blow before you told me the cat died. You could have told me that the cat was on the roof, fell off the roof, was seriously injured, and, sad to say, died from its injuries." A few weeks later, the younger man calls his brother to report: "Grandma was on the roof . . ." Forget the roof. When you need to tell a child (or anyone, for that matter) about a death, it is best to go with the headline first ("Your brother was killed . . ."), then the facts ("in a car accident on his way home from college last night"), and delete the editorials. Euphemisms, platitudes, and proclamations are attempts to soften the blow. They don't. In reality, they are hidden stashes of denial and avoidance that mask the truth, keep death obscured, and in vain try to smother the pain. They are covers for our own unease with death and our self-doubt about what we claim to be feeling. They only increase confusion and fear.

Euphemisms:
"Daddy passed away."
"Your brother has gone to the light."
"Grandpa passed over."
"She expired."
"We've lost your sister."
"Your aunt is sleeping."

"He's gone to his final resting place."
"Your cousin left us."
"He's moved on."
"She's with your grandmother."
"Your little brother has gone on a long journey."
"He checked out."
"God wanted your mother."

Platitudes:

"It was for the best."
"God only takes the best."
"Only the good die young."
"You are only given what you can handle."
"Think of what you have to be thankful for."

Proclamations:

"It's God's will."
"You should be glad that your brother is no longer in pain."
"She's in a better place than we are."

We might feel we need to say something more, even if we have nothing to say. No amount of talking, theologizing, rationalizing, or confronting will ease the pain. Your children need gentle honesty and caring silence. Stick with the headlines and facts, then be present to hold your children, cry with them, and answer any questions they might have. They might be shocked and unable to do anything but cry, or too shocked even to cry. They probably won't be interested in lots of details; just let them know you will be there for them if they have any questions. Assure them, "We are going to get through this together. Sometimes I'll cry, sometimes you'll cry, sometimes we will cry together. We'll even laugh and talk about the good times we had with Dad."

If there ever was a time for the TAO of Family, it is now.

Time spent with our children giving them affection and optimism, even as our own lives are thrown into chaos and confusion, affirms that both grief and joy are vital and inevitable parts of life. Six critical life messages—the ones we often use to help our children get through the everyday ups and downs of life—now become our life raft as we are all swept up in this "tidal wave of grief."

- I believe in you.
- I trust in you.
- I know you can handle this.
- You are listened to.
- You are cared for.
- You are very important to me.

The TAO of Family combined with these six messages becomes the TAO of Mourning.

The TAO of Mourning

There is no despair so absolute as that which comes with the first moments of our first great sorrow, when we have not yet known what it is to have suffered and be healed, to have despaired and recovered hope.

—George Eliot, *Adam Bede*

Since a family has members of different ages, at different stages of physical, emotional, and intellectual development, who have different relationships within and outside the family, no death the family faces together will have the same effect on everyone. There are several factors that will greatly influence children's grieving:

1. Who died and what relationship that person had to the child

If Grandpa lived in the same house or was far away; if it was an older sibling, a twin, or an unborn, much-anticipated baby sister; if it was Mom or Dad, an aunt or uncle, a close friend who's always lived next door, the schoolyard bully, or the new kid in class who was just becoming a friend—it is the relationship already established, as well as the loss of the potential in that relationship, that will influence each mourner.

When a brother or sister dies, children often feel anger toward their parents for not preventing the death or not being honest about the severity of the sibling's condition. They might feel guilty for still being alive or feel that the "good brother" died. They might feel guilty for having any negative thoughts about him and be afraid to say anything that might upset their grieving parents. They might try to take their "perfect sister's" place. They might feel that they can't compete with their dead sibling's idealized memory. They become acutely aware of their own mortality, afraid that they might be next. They might feel afraid, alone, isolated, ignored.

A seventy-year-old friend, Dee, speaking of the death of her twelve-year-old brother when she herself was six, talked about mourning the loss of her mother more than the death of her brother. Lost in her own grief, her mother was unable to be present for her two living daughters. "It was my fourteen-year-old sister who brought my mother back. She told her that our brother had died, but we were still alive and needed her now."

If there are two siblings and one dies, the other becomes an "only" child. That sibling often becomes the silent mourner as people try to console the parents. At the funeral of his seventeen-year-old brother, twenty-year-old Jeff was in a daze as friends of his parents told him to be strong for his parents' sake. He finally blurted out, "I don't have to be strong for them.

They have one another and they have me. I don't have my brother."

When a parent dies, children try to cling to memories with an ever-present fear of "forgetting" the parent. They long to know the person they never really knew. Young children have a strong wish to be with their dead parent, and, using magical thinking, think they can make it happen. A four-year-old told his grieving father not to worry, Mommy was coming back tonight because he told her on the phone to come back. They might feel helpless and abandoned at each major passage in their lives: "If only Dad were here to see this," "If only Mom could see my new baby." They must come to grips with a major loss in their lives and the reality of death, when what they are counting on is an ever-present parent giving unconditional love. Comedian Rosie O'Donnell, whose mother died at thirty-eight of cancer, remarked, "I knew at ten years old that you had a limited run."

2. Manner or cause of the death

The anticipated death of a very ill brother is different from his sudden, accidental death. Any death brings with it suffering, but some deaths bring a compounded suffering, complicated by the cause of the death (suicide, murder, accident), the manner of death (violent, preventable), the people involved with the death (parent, sibling), the fact that it is invisible (perinatal death), the absence of a body or a body recovered much later (plane crash, avalanche), the death's being a part of other destruction (natural disaster, fire), or a compounded tragedy (murder and suicide together).

These are the hard deaths to grieve, because grieving gets compounded by the many emotions that collide—rage and sorrow, relief and pain, frustration and hurt. As well, the mourners of such deaths often must deal with rumor and innuendo, trips to the police station, scrutiny by the news media, their private lives and personal grief made public. The grieving after a murder or suicide is compounded even more by the five S's that are often attached to such deaths: stigma, shame, secrets, silence, and sin.

These same five are attached to some terminal illnesses as well. When a person is dying of AIDS, the whole family might feel isolated and shunned, unable to grieve the death of their loved one openly for fear of the judgmental reactions or condemnation of others, enveloping themselves in shame and silence. Even worse, members of the family might be the ones to do the shunning, judging, or condemning, denying themselves and the dying person the opportunity to share in the passages of grief and express their love. The Reverend Susan Phifer of East United Methodist Church, speaking at the United Methodist General Conference in Denver on April 19, 1996, said, "I do not profess to understand all of the complexities of human sexuality . . . but for myself, when I stand before my Creator and he asks, 'How did you treat my children?' I want to err on the side of love and grace. I don't want to err on the side of a self-righteous judge." Love and grace are what any person needs when dying.

3. Communication skills of the family

Secrets, white lies, and hidden agendas create barriers to mourning. Repression of feelings or inappropriate acting-out inhibit the ability of family members to support one another. Children need honest and truthful explanations, the opportunity to cry and to see others crying openly. They need to be able to express their anger, hurt, disbelief, and fears, knowing that someone is listening and providing the needed support and appropriate boundaries for these expressions. They need to be able to tell their own story of loss, sometimes over and over again, so they can begin to accept what has happened and to heal.

4. The history of loss and death

With a first death experience come shock, sorrow, and grief, without the comfort of knowing what it is like to heal and to find joy and hope again. Simultaneous multiple deaths, death that reminds you of an earlier death, deaths in quick sequence, unexpected death when in anticipation of another death—all these factors can compound mourning.

5. The developmental level of the child

Children of all ages can experience loss and grief. The ability to express the grief and understand death is in large part dependent on the developmental level of each child. A toddler asks when his dad will be done with being dead, his eight-year-old sister asks why, and their teenage brother asks about life after death.

As much as we want to protect our children from chaos and loss, when a death does occur, we cannot hide the event, even from small children. They are perceptive and aware of our feelings of sadness, anger, and hurt. They hear our muted conversations; they see us crying. They know something has happened, and they try to put together the pieces of the puzzle, using imaginary pieces if the real ones are not given to them.

If they are allowed and encouraged to mourn and supported in their grief by caring adults who offer them their time, affection, and sense of optimism, children of all ages learn that they, too, are capable of handling powerful emotions, have the ability to work through their grief, and can help others in their grieving. In fact, it is often a child who, unabashedly and without concern for appearances, reaches out to comfort and reassure others who are grieving. After the crash into the ocean of Swissair Flight 111, a photographer snapped a picture of a small child handing a flower to her aunt, who was standing at the debris-strewn shoreline weeping for her dead husband. To be included in the mourning after this catastrophic accident not only helps the child understand the grieving happening around her, it also allows her to be an important member of her family's mourning circle, and thus its healing circle.

We can give them the necessary facts, ways to express their grief, and our assurance that through all of this chaos they will be taken care of and loved. Just as we could not prevent the death, we cannot shield our children from the loss. However, using the TAO of Mourning, we can buffer our children of all ages from unnecessary and potentially devastating consequences of the loss.

Ages and Stages

Unborn child

A child in the womb is aware of its mother's grief, sadness, and depression and the expressions of these feelings and might feel distressed or agitated. Because the two of you are so intimately connected, your physiological responses to grief and mourning affect your entire body and therefore influence the environment of your unborn child.

TAO of Mourning: Provide loving care to yourself, eat as well as possible, and rest as often as possible. Let others help. Tell your obstetrician about the death. For your sake and your unborn child's, take breaks from the mourning environment, go for a walk, find a quiet space, put on soft music, and make time to enjoy the new life in your womb, a welcome reminder that life does go on.

Infants

Infants function in the present. They are tuned in to their environment and aware of presence, sudden change in physical and emotional climate, and absence. The death of the mother is an infant's greatest loss, since the mother is the person most depended on for comfort, security, and stability. The infant might respond to the death of the mother with irritability, crying, fitful or prolonged sleep, and marked changes in eating habits. If the mother is grieving the death of a loved one, the infant will sense the grief and respond to it with agitation, possible changes in bowel activity, spitting up.

TAO of Mourning: Provide loving, consistent care. Respond to the infant's needs with gentle touching, talking, and singing. Avoid rigid, angry, agitated responses. The first three parts of the six critical life messages are for you and the last three for the child. Reminding yourself that you believe you can make it through this can help strengthen your resolve to listen to and care for your child during this chaotic time. It is the little things you do for her that will affirm her importance in your life.

Toddlers

Toddlers are actively involved in doing. They are learning that it is safe and wonderful to explore, yet they want to be able to get their parents' attention at will. They are curious, testing all of the senses; they express a wide range of emotions. They are tuned in to and can "read" other people's moods. Language is better understood than expressed.

Toddlers can't conceptualize death but will express sadness even when a pet dies. They need an explanation that the body has stopped working and will not work again. They will experience a profound sense of loss at the death of their mother. Toddlers grieve in spurts of five to ten minutes, throw temper tantrums, rock themselves for comfort, and often revert to earlier behaviors, such as thumb-sucking.

TAO of Mourning: As with infants, provide loving and consistent care. Respond to their needs with gentle touching, talking, and singing. Talk about why you are sad, and be comfortable crying in front of them and with them. Assure them: "I am very sad right now because I miss your grandpa very, very much—and I will make you that peanut butter sandwich you asked for. We will go for our usual walk to the park and play on the swings." In taking the time to do these ordinary, everyday activities while you are grieving, you help your children know that life does go on. You will be there for them even in the midst of your own grief.

Preschoolers

Preschoolers are busy establishing an individual identity, learning new motor and linguistic skills, and figuring out roles and power relationships. They look to siblings and adults as models for their own expression of emotions. They ask questions about their bodies and the world around them. They may ask *why* people have to die. They are trying to separate fantasy from reality.

Preschoolers tend to see death as temporary, a journey from which they can return or a long sleep from which they can awaken. They might nod solemnly when told of their father's death and still expect him to show up for dinner. They feel loss, experience a wide range of strongly felt emotions, and grieve. Even though they might be very verbal, their ability to put into words their tumultuous feelings will be lagging behind their body's experience and expression of those emotions.

When told of a death, preschoolers might appear sad, bewildered, or ambivalent and act out those feelings through play. They might speak their feelings to an imaginary friend. When watching movies such as *The Lion King,* they might assume the role of the grieving Simba and actually hunch their shoulders and mope, imitating in facial expression and body posture the sad and bewildered lion son. They might use bullying and aggression to cover fears, anxiety, and sadness. They will cling to a parent, hang on for security, be afraid that Mom or Dad will leave, too. They will run outside to play and come back to the screen door often just to make sure that Mom or Dad is still there. Preschoolers often attach themselves to a person who looks like or acts like the person who died and call that person by the deceased's name.

TAO of Mourning: As with younger children, provide the loving care, affection, and attention they need. Keep their daily activities as routine as possible. This is not the time to introduce new activities or radically change their schedule. Now is the time to discuss death in more detail, adding the headlines and facts about the death as well as the four attributes of death:

1. It has a specific cause. *(Grandpa was very, very, very ill.)*
2. It involves cessation of body functions. *(He can't move, can't feel, can't breathe, can't grow, and is not just sleeping.)*
3. It is irreversible. *(Nothing we do can make him alive again.)*
4. It is universal. *(Leaves die and fall off the tree, plants die, animals die, and Grandpa is dead.)*

Be available to answer the same questions over and over again. Preschoolers use "magical thinking," believing their thoughts, words, and actions have great power. Reassure them again and again that nothing they thought, said, or did caused the death, and that nothing any of us say or do will bring the dead person back to life.

Five- to Nine-Year-Olds

Youngsters from five to nine tend to listen in order to collect information; compare; test; disagree with adults and peers; set, break, and change rules. They challenge parents' values, argue, and hassle. They are openly affectionate at times, self-contained at other times. They can separate reality from fantasy and can use intuition to help them decide what to do.

They now understand the reality of death, that it is irreversible and can happen to people close to them. The fear of abandonment is expressed verbally at this age. If death can happen to one person, it can happen to others, including themselves. Concern for survival of self includes how the death will affect them personally. They worry about their own health and the health of surviving relatives. Feeling vulnerable, they might deny that the death happened or that it hurts them in any way, acting tough to hide the pain. They will play with great passion, trying to drive the pain away. They might search room to room, out in the yard, at the ballpark for the deceased. They often idealize the deceased and might get attached to articles of clothing or objects belonging to him or her.

TAO of Mourning: Kids at this age will often ask many specific questions about the death, the circumstances surrounding it, and the biological aspects of it. Going over the four attributes of death with them will help ease their fears: Death won't come suddenly in the night while they are sleeping; Mom won't die just because she left to get the groceries. They need to be encouraged to grieve openly—if not in public, then at least in the safety of their own family or with a trusted family member or friend. Drawing, painting, or molding clay can help them express feelings that they find difficult or are afraid to share.

Talk with them about the passages of grief and what they can expect in those passages. Youngsters at this age can be helped through their grief by helping others who are also grieving. Be cautious not to overwhelm them with all of your grief or with all of your knowledge.

Answer questions honestly. As Sarah was tucking in Seth, her eight-year-old son, he asked what Grandpa looked like. Thinking he was already forgetting, Sarah grabbed a photo of Seth and his grandfather, taken a few months before he died. Seth shook his head, said he knew what Grandpa *used* to look like. What he really wanted to know was if Grandpa was a skeleton yet. A quick explanation of embalming fluids and the process of the body tissues breaking down was sufficient for the moment. The next moment the discussion turned to the photo of Seth and his grandfather sharing a fishing pole.

Preteens

Unable to be adults, unable to be children, and wanting to be both, preteens are shier about crying or sitting on someone's lap, even though they might want desperately to do just that. They also yearn for all the independence that comes with being a teen, to be left alone to put the pieces of their grief puzzle together by themselves.

They know that death is permanent and that it might come earlier than expected. They want all the details of the death and want the facts about how the death will affect their everyday life. At the same time, they will often deny that it affects them at all. They might sulk, say they don't feel anything, say they don't care, in an attempt to bury their intense grief and their unsettling fears about their own mortality. They might make exaggerated attempts to help others while being superstrong themselves, as though this display of strength will keep them from being tainted by the loss. They fear being different from their peers.

Preteens may complain about headaches, stomachaches, inability to sleep. They might express anger at the deceased for dying or express guilt about what they would have, could have, might have done to prevent the death. They will often replay the events leading up to the death in a futile attempt to construct a different ending. They might experience nightmares related to the death or wear clothing belonging to the deceased, listen to music the deceased liked, make a montage of photos of the deceased and themselves together.

TAO of Mourning: Preteens need everything the younger kids need: your time, your affection, and your expressions of optimism. Since preteens need their parents to be mentors, it is important that you express your feelings with them and show them how you are doing everything you can at this point to get things back on track. They need reassurance that life will go on, that you all will make it through this trying time, and that eventually the piercing grief and intense suffering will give way to sadness, peace, and an enthusiasm for life. As well, preteens need opportunities to spend time with their peers to mourn, to laugh, to have fun. Their peers can help them move through the passages of grief and offer them a respite from the pain.

It is tempting to criticize preteens when they revert to younger behavior or don't express their grief at the appropriate time or in the appropriate way. Bite your tongue. In your optimism, you have to believe that they are doing the best they can right now. Keep going over the six critical life messages in your head and then find small ways to express them to your preteen. Do, however, confront attempts to idealize the deceased. Help preteens remember the deceased with his or her many warts as well as beauty marks.

Helping with the rituals of mourning, creating their own way to say good-bye, retelling the story of their relationship with the deceased, and commemorating that relationship will help make the death a reality that preteens can face with courage.

Adolescents

Take all the possible responses of the preteen, magnify them with the reasoning ability of an adult and the erratic emotional states of a teenager, and you have the common characteristics of adolescents facing the death of a loved one.

Just as preteens are in a transition from being children to teens, adolescents are on the cusp of being adults. Emerging as separate, independent persons with their own identity and values, they are able to understand the concept of death and want immediate answers to the possible implications (including financial ones) of a particular death in their own lives.

Understanding does not mean accepting. The adolescent will often talk about the death abstractly, use gallows humor to cope, idealize the dead person in the morning, rip him apart verbally in the afternoon, in the evening swear never to mention the deceased's name again, and at bedtime want to play a favorite song of the deceased, only to listen to the first line and then turn it off in despair. Emotions are in turmoil, moods change abruptly, swinging erratically from sadness to anger, giddiness to anguish. Adolescents might demand all the independence that the preteen yearns for and retreat into their own bedrooms for hours on end to sleep, cry, or pore over photo albums. They often share their thoughts and feelings with close friends. Some retreat into a mind-numbing depression or use drugs, alcohol, or food to drown the pain. They might question life's purpose and meaning, rage against the inequities and unfairness in every area of their lives, and swear never to get close to anyone again.

TAO of Mourning: Adolescents need everything that preteens need from you. As well, they need opportunities to help in the decision-making and planning for ceremonies, rituals, meals, and activities associated with the burial, memorial services, and commemorative activities. Speaking at Grandma's funeral service might give them the opportunity to put their grief into the context of the wonderful relationship they had with Grandma and the memories they will always carry with them. They can be encouraged to come up with their own meaningful rituals. The songfest in memory of a dead classmate might not be your idea of a memorial, but it is meaningful to them.

Be aware that adolescents might feel a need to be "strong" for you if they see you falling apart. This strong front will only delay and perhaps prolong their grieving. Help them name their grief and respect their emotions. In grieving there are no *shoulds* and *oughts*. Be aware of their expressed desire to be with the person who died. It is normal to "wish I could be with him." Don't be alarmed, just listen and show your understanding.

Do be alarmed and get your teen help if he is preoccupied with the wish to be with the person who has died and talks about *when* and *how* he intends to make the wish become a reality. Be alert to signs of deep depression, severe fatigue, alcohol or drug use, overeating or undereating. Take any talk of suicide seriously. Far better to look foolish than to have lasting regret.

Life is in your hands. You can select joy if you want to or you can find despair everywhere you look.

—Leo Buscaglia, *Born for Love*

When Grieving Is No Longer Good Mourning

The time at length arrives, when grief is rather an indulgence than a necessity and the smile that plays upon the lips, although it may be deemed a sacrilege, is not banished.

—Mary Shelley, *Frankenstein*

Sometimes grief is blocked, diverted, or buried. The following is a checklist of warning signs that your child might be stuck in grief and need professional help to get through mourning. All children will exhibit some of these signs as they grieve. It is the frequency, intensity, and persistence of these behaviors that would indicate a need for concern.

1. Acting much younger for an extended period of time.
2. Excessive and prolonged crying bouts.
3. Inability to sleep or need for excessive sleep.

4. Nightmares or night terrors.

5. Loss of appetite.

6. Extended period of depression in which child loses interest in friends, daily activities, and events; putting a negative spin on events.

7. Truancy or a sharp drop in school performance and grades.

8. Prolonged fear of being alone.

9. Persistent idealization of the dead person.

10. Excessively imitating the dead person.

11. Repeatedly stating the wish to be with the dead person.

12. Clinging to the past and refusing to think positively about the future.

13. Talking about the dead person in the present tense.

14. Overvaluing or clinging to possessions of the dead person.

15. Frequent physical complaints, illness, headaches, stomachaches.

16. Detachment and pulling away from efforts at consolation.

17. Avoidance of any activities that might be a reminder of the dead person.

Sadness can coexist with peace, hope, and joy. Depression cannot.

—**Andrea Gambill**

Suicide

Suicides are messy deaths: there is nothing neat about them. The lives of those who are left behind have been shattered into thousands of tiny fragments, and we do not know how to begin cleaning up the devastating damage. Our loved ones have departed by their own will; even though they knew that they were planning to leave us forever, they did not give us the opportunity to bid them Godspeed.

—**Carla Fine, *No Time to Say Goodbye***

Suicide is the eighth leading cause of death today, and the third leading cause of death among young people aged fourteen to thirty-five. I have included a separate section on suicide because it, more than any other violent death, could touch your family's life and because it has many of the hallmarks of other complicated deaths: It is sudden, unexpected, violent, messy, intrusive, irrational, with disorienting and devastating repercussions, leaving mourners feeling powerless, inadequate, hurt, and betrayed.

I knew when my mother phoned while I was away on a road trip that she wasn't just calling to discuss the weather. Mercifully she blurted out the news, without any warm-up or attempt to soften the blow: "Jim is dead." My friend since second grade, the king of the fifth-grade Valentine's party when I was the queen, my date for the senior prom, the one who wrote beautiful poetry, the one who took me to *The Sound of Music* days after I left the convent, the one who sang "Let It Be" at my wedding, the award-winning author, the university professor, the husband of Teresa, the son of Betty, the brother, the uncle, the friend—he had completed his final book, dedicated it to his mother, then wrote two notes and hanged himself from a tree in the mountains he loved.

Our town newspaper reported only his death and a bit about his life. The big-city newspaper gave the gory details—details that linger but yield no answers. Suicide always leaves more questions than answers. The lines from Voltaire kept running through my head: "The man who, in a fit of melancholy, kills himself today, would have wished to have lived had he waited a week." Just to have that one week. But to someone who is depressed, a day is too long. No amount of success or good news could pull him out of his depression. Had he simply run out of options to ease his pain?

Suicides *are* messy deaths. The people who found the body will remember the scene for a lifetime. Jim's wife and mother got written notes that offered little if any comfort and certainly no explanation. Suicide notes are rarely coherent; they are usually written by someone in pain and confusion. The violent nature of the death leaves in its wake shock and recurrent nightmares. Coroners often report that, even in death, the body of a person who has killed himself seems physically tormented and not at peace.

When anyone commits suicide, the mourners are overwhelmed by the irrationality and irreversible nature of the loss. They often feel disoriented and a bit crazy themselves. Amid the personal chaos, the nature of the death forces the private to become public, with police reports, pictures of the scene, restrictions on the burial rituals, suicide notes documented and scrutinized for clues that usually don't exist, questions and more questions. How can anyone be expected to give rational answers when overwhelmed with grief and simultaneously filled with rage? Besides, what good are any answers? There are no answers that will change the outcome. The opportunities to reverse the outcome don't exist, except in the mind of the mourner, forever looping over and over the days or moments before, trying desperately to create a new ending.

Suicide, a seemingly personal act of the will, becomes a public, criminal act identified as a homicide until proven otherwise, or irrefutably identified as a suicide, but a *criminal* suicide nonetheless, or a suspicious death that was perhaps a well-planned "accident." Taking someone's life, even your own, is against the law in most places. This only multiplies the problems the mourners will need to handle in relation to wills, life insurance policies, creditors, and even the burial itself. The public nature of the act seems to give some onlookers the perceived "right" to ask any and every intrusive and curious questions they would never ask about another person's life, let alone ask in the wake of a death. Others openly condemn or censure the act, giving no thought to the mourners' grief.

Some religions still place restrictions on the burial of a suicide victim. Thankfully, many do not. Recognizing that the person who commits such an act is often seriously emotionally or mentally ill, "out of his mind," and often desperate, many clergy are more concerned with the well-being of the family of the suicide victim. The act of suicide is not denied, but neither is the lifetime that came before the act itself. The person has a name, a history, family and friends who grieve the death as profoundly as they rage at the act. They need comfort and compassion.

One aspect of suicide that sets it apart from other deaths is that the mourners must grieve for the very person who has taken the life of that person. The victim and the villain are one and the same. The decision to die and the no longer being alive occur in the same time and space, to magnify the grief. The mourners feel a deliberate abandonment and rejection, while at the same time they feel an anger and rage. The "why?"s and the "what if?"s combine with "How could she do this to us?" and "Why couldn't she come to us?" Mourners can't just experience the piercing grief of good-bye; they get hammered by these questions and hammered by the reality that they belong to a different group of mourners than any they have ever been a part of before.

The five S's, unbidden and unwelcome, arrive to sit squarely in the living room, daring anyone to speak their names and shatter their presence. It is only in speaking their names and shattering their presence that the mourners can get on with their grieving. Stigma, shame, secrets, silence, and sin can remain only if those of us who grieve do so silently, feeling shame, afraid to speak the seemingly unspeakable word "suicide" and give it its due, but no more, in the life story of the one who has died. We need to speak the person's name, tell his entire life story, and laugh. Laughter helps us to see life's ironies and recognize the larger whole of the story.

When it is a child or a teenager who kills herself, all of the above become even more complicated and compounded by the profound grief of a child's dying, added to the guilt of our somehow not preventing the death, the helplessness at not being able to keep her alive, the accusations of lousy parenting ("How could they miss her depression?"), the regret of not being able to say good-bye, and the incalculable loss of the future, with its possibility of good memories and a good relationship. With teenage suicide, families are often left with lasting memories of the turbulence that marked the few years or months before the death, with no opportunities for resolution or reconciliation.

Although there are ways we can buffer children to keep them from killing themselves, there is no way to inoculate them totally from it. We can create a home environment in which our children grow up knowing that their feelings are accepted, their ideas count, their basic needs are met, and their mistakes are seen as learning opportunities. We can provide the structure to help them flesh out a sense of their true selves and give them the tools necessary to help them solve the myriad problems they will face. We can listen to their cries for help— not ignore them, laugh at them, or dismiss them as foolish. We can offer them our time, affection, and a sense of optimism. We can give them our love and show we care. In the end that is all we can do. As one father said about his son's suicide, "I came to understand that while it might be possible to help someone whose fear is death, there are no guarantees for a person whose fear is life." Carla Fine wrote in her own memoirs about the suicide death of her husband, "I had to accept that, ultimately, it was Harry's own choice to kill himself. All I can do is disagree with the decision."

It is those we live with and love and should know who elude us. Now nearly all those I loved and did not understand when I was young are dead, but I still reach out to them. . . . Eventually all things merge into one, and a river runs through it. The river was cut by the world's great flood and runs over the rocks from the basement of time. On some rocks are timeless raindrops. Under the rocks are the words, and some of the words are theirs. I am haunted by the waters.

—Norman Maclean, *A River Runs Through It*

Chapter 3
• • • • • • •

When Illness Strikes

And you will accept the seasons of your heart, even as you have always accepted the
seasons that pass over your fields.

—Kahlil Gibran, *The Prophet*

The journey of the heart that each one of us takes as we face the
death of someone we love is the same journey we will take when
someone in our family is diagnosed with an acute illness, a chronic illness, or a disability. The three passages that we circled through—the
piercing grief of good-bye, intense sorrow as we reorganize our life, sadness that shares space with a quiet joy and a gentle peace—will become
well-worn paths we will travel over and over again. Occasionally we'll
meet our children going in the opposite direction; sometimes we'll
travel with them hand in hand. And sometimes we'll be ahead of them,
wishing they would catch up, only to find out they have already been
there, done that, are ready to get off and end their journey as we must
continue in ours.

An **acute illness**—such as chicken pox, meningitis, strep, ear infection, pneumonia, or appendicitis—can cause exceptional stress on a
family for a short period of time. By definition, acute illness is sudden,
brief, and severe. It can be a wake-up call to remind us of the gift of
good health that we take for granted when we are well. It can help us
put petty problems into perspective. Acute illness can also bring us to

our knees in grief if our child dies or is seriously handicapped as a result of complications.

Jonathan was a spirited child, full of all the wonder and skepticism of a normal two-year-old, gleefully throwing his sandwich crust on the floor at lunchtime. By four o'clock he had a fever, by eight he was hooked up to machines, and by four in the morning he was dead from a virus. Down the hall was another two-year-old depending on life-support machines as her organs began to fail one by one. By seven the next day, Melissa was responding to her mother, and three weeks later she left the hospital, smiling and waving to the doctors and nurses. A third child, Samuel, left the same ward two weeks later to begin rehabilitation in an attempt to relearn what the virus stole from him. All three had been healthy two-year-olds; each one's family was irrevocably changed by their child's acute illness; each one's journey of the heart was very different from that of the others.

Such is the nature of the passages of grief through any illness or disability. Know that the passages you will travel might loop backward before going forward, as they did for Jonathan and his parents: the piercing grief and the intense sorrow related to his illness circled back to the piercing grief and intense sorrow of mourning his death. It would be a couple of years before his parents would know the gentle peace and quiet joy of the third passage. For Melissa's family, it took a short time to get through the first two passages and a lifetime for the third. Samuel's family circled around all three many times, feeling the three occasionally colliding. For his family, the piercing grief of the handicap and the joy of life were present in the heart at the same time.

The same can be said for a **chronic illness,** one that is of long duration and/or recurring. A chronic illness can cause minor or major stress on a family for a long time, sometimes for a lifetime. Its cause is not always apparent: a real-life tragedy with no known villain. It is often invisible. It can radically change family routines, rituals, and traditions that were in place long before the illness became a permanent fixture in the home.

Chronic illness is more likely than an acute illness to affect your child's psychological and social development. Extended absences from school and extended trips to the hospital can exacerbate moodiness and a sense of social isolation. Your child's role in the family might change markedly, especially if the chronic illness results in a permanent disability. Six of the most common chronic illnesses that affect children today are asthma, severe allergies, arthritis, diabetes, seizure disorders, and cancer.

An illness can be either acute or chronic depending on its severity or the duration of treatment. Cancer can be an acute illness that is severe, short in duration, and deadly, or severe, short in duration, and curable. It can also be a chronic illness, such as leukemia, requiring long-term treatment and the possibility of recurrence.

Both acute illness and chronic illness can cause a **mental** or **physical disability** that can have a small impact on a child's life or a large impact on the entire family's life. Accidents and mayhem can also cause acute trauma and/or permanent disability.

Chaos and loss don't come to us in sequence, one at a time, on simple terms we can understand, wrap themselves up, and go away before we are faced with another simple, easy-to-solve loss. If a child is born with multiple handicaps, often one or two show up immediately or shortly after birth, while the others show up later. Some might actually be the result of medical intervention for one of the first handicaps—an infection around a shunt to alleviate pressure on the brain, for example, can lead to a life-threatening illness.

A child might grow normally, physically and mentally, until he develops an innocuous fever at two years old and then begins displaying autistic behaviors. These behaviors can range from mild to severe, with little known about cause or treatment and no known cure. Another child can be diagnosed at age ten with leukemia after complaining of pain in her legs after a soccer match, the disease striking like a bolt out of the blue.

A preteen suffers a severe head injury and makes a remarkable recovery with only the emotive area of the brain left seriously damaged. Looking and acting physically normal, she is unable to sustain personal friendships and is dangerously aggressive. Another suffers a similar injury, along with crushed vertebrae, and is paralyzed from the neck down. With his indomitable spirit and determination, he grows up to teach elementary school.

Two teens in the same school district are diagnosed with tumors: one benign, one malignant. The benign tumor is in the brain and is inoperable. The teenager dies before he can complete his senior year. The other tumor is a rare radiation-induced follicular carcinoma in the thyroid. The tumor is removed; radiation treatment eliminates any trace cells; the teenager finishes high school, graduates from college, and goes off to Hollywood to pursue her dream of being a stuntwoman.

And in this life there is no rule that says two children in the same family cannot be stricken with serious illnesses at the same time or in succession. Illness and fairness rarely show up in the same sentence.

During her senior year of high school, our daughter Anna was diagnosed with an acute illness, Lyme disease, that if left undetected or untreated could have resulted in a severe chronic disease. Our other daughter, Maria, was diagnosed her senior year with cancer, and their brother, Joseph, refused to see a doctor his senior year, ruling out for him any possibility of getting diagnosed with any disease. We were hit hard twice, two years apart. As we were moving comfortably into the third passage with Anna, we found ourselves back at the first with Maria. Anna proved to be a gentle guide for Maria, as Maria was now the one who needed a hand, having lent hers to Anna two years before. Joseph kept us painfully aware of the grief a sibling goes through, sometimes in companionship with his sisters and at other times pulling us all along into the respite of the third passage.

The Passages of Grief

Look well into thyself; there is a source of strength which will always spring up if thou wilt look there.

—Marcus Aurelius

1. Piercing grief of good-bye

When a diagnosis has confirmed what our mind and heart already suspected or hoped was not true, we are thrown into the shock of this first passage. We might try to deny the truth; it does no good. We might rage at the doctors, at the gods, at life itself. We are in a state of disorder and disbelief. We numbly go through the routines of the day, not remembering how we got from one place to the next. We look in medical journals to find the exception that does not exist for our child, the cure that has not yet been found. Hopes and dreams are dashed, and it is too soon to be optimistic about new hopes and new dreams.

The good-bye of death is final. This good-bye keeps on going, like a nightmare that relentlessly intrudes and refuses to be chased away with wishes or bargains. Our child's body or mind or both are affected by disease or disability, changed forever in an instant by an accident, a virus, a genetic defect. Or perhaps the change is an insidiously slow deterioration from a chronic disease or a malignancy. It matters not the why or the how—it is the reality of the illness that hurts. The world seems unstable, unpredictable, and unmerciful.

In death there are rituals and routines that help us put closure on the event. This is not true with an illness. There are no rituals, only medical routines and therapies, new medicines, and checkups, diagnoses, treatments, and prognoses. We think we have finally steeled ourselves to accept this reality and move on through to the second passage, when a complication sets in—a blood count is low, an illness on top of a disability slows the healing process. We have to find closure before we can move on. Often that closure is the acceptance of the diagnosis. This acceptance is not a resignation or a giving up of the fight; it is the naming of the beast. In that naming we have the power to create our story—granted, with a character we didn't invite and whose lines we would rather not hear.

2. Intense sorrow as we reorganize our life

As we begin our new story, we move into a time of deep sorrow. The full impact of the diagnosis and its implications hits us. We have little energy for the ordinary tasks of life, and now we must add therapies, medicine, trips to the doctor's office and the hospital. We want to be dormant, to pull into ourselves and escape from the everyday realities of the illness or disability. The "what if?"s are driving us crazy. No longer numb, the mind keeps rolling back time to the moment before the accident, the day before the illness, the night our child was conceived, struggling against reality to re-create a different ending.

But we can pity ourselves for only so long; our bodies and minds need to move out from under this sorrow. It is time to busy ourselves with creating a new identity that includes the illness or disability but is not totally framed by it. It is a time to step back and gain a new perspective. It is also a time when we can reach out to someone else in pain, knowing we have traveled a similar path.

We can find ourselves stuck in this passage if we are unable to get any respite, if we are in an unending cycle of intensive caregiving. Our child will be stuck in this passage if the illness or disability consumes her every waking moment. Sometimes the treatment or therapy needs to take a backseat to everyday events or a trip to the park.

3. Sadness that shares space with a quiet joy and a gentle peace as we recommit to life ourselves, tempered by the loss and wiser

As we begin to reenergize, we can feel ourselves wanting to laugh and be released from sorrow's burdens. We welcome the new energy and the ability to concentrate on something other than the disability or illness. We don't feel bad that we feel good. We know that we have been through the valley of darkness and have come out on the other side, changed but very much alive. It is not a matter of getting over the losses that the disability or illness creates. It is knowing that we can get through the pain and the grief. Only people who have known a great loss can really know the quiet joy and gentle peace of this passage.

Knowing we can get through the pain and the grief does not keep us from returning to the first and second passages when we are hit by a new loss or when an old loss comes back to haunt us. It's just that once we have been through the third, we know it is a place we can find again.

Zackary, a tall sixteen-year-old, asks for a blue truck for his birthday. He wants the real thing but must settle for a model because he was born with Down's syndrome and will never be able to get a driver's license. His pain is as palpable as his mother's as they both once again feel the piercing grief and the intense sorrow that have been their ever-present companions since the day he was born. Most of the time these companions sit in the backseat, along for the ride, but every once in a while they grab the wheel to remind both Mom and Zackary that some hopes were dashed forever, some dreams will never come true. So Mom and Zackary dream different dreams.

This journey of the heart, the circle that directs us back to the present moment with all of its possibilities, is also a journey of the mind and of the soul. The Maori of New Zealand have a saying: "The mind, the heart, and the soul; each one has its own perspective." As we move through the passages, we don't find answers—there are no good ones when it comes to a disability or an illness. What we need to find is a new perspective, a new way of seeing.

When one door of happiness closes, another opens; but often we look so long at the closed door that we do not see the one which has been opened for us.

—**Helen Keller**

Diagnosis, Treatment, Prognosis

There are three kinds of lies: lies, damned lies, and statistics.

—**Mark Twain, *Autobiography***

When you have been hit with the news of your child's acute illness, chronic illness, or disability, the mind grasps for some meaning, some label to put on the beast in an effort to understand it and understand what it is doing to your child. Knowing which questions to ask can help you in this time of great confusion. The questions fall under three categories: diagnosis, treatment, and prognosis, or DTP.

As a family we got lots of practice with these questions through our experiences with a minor injury (Joseph's collarbone), an acute illness (Anna's Lyme disease), and an acute/chronic illness (Maria's cancer). There are things we did, wished we had done, didn't do, and hope we never ever have to do again.

For Joe the diagnosis was quick: a broken collarbone. The treatment was noninvasive, simple, and low-tech: a shoulder brace, some rehabilitative exercises, and pain pills for a short time. The prognosis was excellent: a small bump on the collarbone as a reminder to take the leaps on a snowboard with a little less abandon and a little more common sense.

Anna's diagnosis was more difficult and took longer, with a

range of possibilities, from a brain tumor to multiple sclerosis to Lyme disease (which can mimic both the others). Treatment was more invasive and more high-tech: brain scans, spinal tap, neurological exam, a regime of potent antibiotics over a four-week period, self-administered shots, and lots of blood tests. The prognosis was excellent, with only minimal concern for undetected consequences from the disease.

Maria's diagnosis took less time than Anna's, more time than Joe's, and took us in more directions—down into a few valleys and over a mountain of possibilities. Her treatment was high-tech and invasive. Her prognosis was excellent, with the nagging concern of future cancer—a risk no greater than that faced by most of the population, and certainly a greater chance of detection with all the monitoring that is now a part of her life.

The DTP model fit for all of the kids. But diagnosis, treatment, and prognosis do not always follow one another in an orderly sequence. Any number of "what if?"s can change any one of the three and subsequently the other two. For instance, Maria's tentative **diagnosis** required a **treatment** before we had a full **diagnosis** and **prognosis**, which then resulted in additional **treatment,** which in turn affected the long-term **prognosis**. Maria's initial diagnosis was a growth of some kind on the thyroid gland, most likely benign but possibly malignant. The chances were 99 percent that the tumor was benign. The tumor was sent to New York, and the statistics failed us 100 percent. The tumor was a follicular carcinoma, a rare form of radiation-induced cancer. Maria was to join the ranks of the children of Chernobyl, who had similar cancers after Russia's nuclear-reactor accident.

Her prognosis is excellent. Her Synthroid provides a daily reminder of the fragility of life, the complex nature of human hormones, and the wonders of science in a pill. Blood work every six months and high-tech scans every year have become a routine part of her life.

That all said, these words in no way convey the roller-coaster ride of feelings we all went through and still go through today. What for another young adult is just a bad week—a few headaches and a few aspirin—becomes for Maria a brain scan to see if any cancerous cells are lurking in the brain. Our vocabulary now includes words we never knew existed, or care ever to speak again. We know that if lightning can strike once, it can strike again; if one child can have a serious illness, so can the others. If cancer was in the body once, it can return. Life as we knew it changed dramatically. Yours will, too, if you have a child with a serious illness or disability.

As you begin your journey that includes a serious illness or disability in your family, knowing *what* to ask and *when* can help you handle an illness and prevent it from overwhelming you.

Diagnosis

The first step in most medical situations is to make a diagnosis. A proper diagnosis is crucial if the best treatment plan is to be devised. What is the name of the problem? What is its technical name? What does it mean? What are some other possibilities? A speech impediment might be due to a hearing loss, a cleft palate, or a combination of both. A pain in the leg can be from a fall on skis or from leukemia. Often just naming the problem gives parents a life raft to hang on to; they were not being crazy, or overly anxious, or paranoid. Their child's symptoms were real, not imaginary or exaggerated.

Treatment

This step usually follows the diagnosis. Sometimes, though, a variety of treatments are tried before an adequate diagnosis is given. Often children with chronic diseases receive treatment for a seemingly endless number of different ailments, until finally a true diagnosis is made. When there is a name for the whole group of symptoms, a more adequate treatment plan can be found.

The treatment plan is not always simple and straightforward. Sometimes there are options to choose from, such as surgery, or radiation, or a combination of both. Some treatment plans will begin with little if any actual intervention and become increasingly aggressive; others will begin with the most aggressive treatment first. Sometimes only one or the other is needed. Surgery to remove a tumor might be the first choice or the last resort.

Since Maria was three weeks away from her eighteenth birthday when her tumor was discovered, she was actively involved in the whole treatment process—but it was our signature as parents that went on the consent form. If you have a young child, it will be your decision and your signature, but as your child reaches the teen years, usually she will be able to take a more active role in the decisions that need to be made regarding treatment options. In an emergency, you will need to make the decision yourself. If your child is severely mentally handicapped, regardless of his age you will need to make the decisions you feel to be in his best interest.

Prognosis

Prognosis is basically what the outcome of a treatment plan is expected to be. There are no guarantees here, just educated and informed guesses. Occasionally it's not even an educated guess, but more like "I wonder what will happen next?" This means dealing with the good, the bad, and the ugly. Sometimes, to achieve the optimum good, ugly scars need to be a part of the results. When a choice is made to forgo further treatment, the short good time with loved ones might far outweigh the potential for longtime pain with little if any gain. Short-term intensive pain and rehabilitation can open the door to a lifetime of vibrant living. And sometimes it's really a toss-up with no clear "good" choice or "bad" choice, just a personal choice and a leap of faith.

After an automobile accident, one teenager might need only

outpatient rehabilitation for several weeks, with a return to normal activities and routines. Another might need residential care for months and then a return to normal activities and routines with minor adjustments. Still another might need residential care for the rest of his life.

It will be your job as a parent to become the investigator, insurance navigator, and advocate for your child. That means educating yourself, getting good advice, and figuring out a way to make the medical system work for you. Don't be afraid to ask for specifics. If you don't understand what is being said, ask for it to be repeated or rephrased. There are no foolish questions.

Take notes. You think you will remember what has been said to you, but when you leave the doctor's office, the pain and anguish you stuffed to get through the appointment might overwhelm you and make your mind go blank. Keep a medical journal of doctors, specialists, appointments, diagnoses, treatments, therapies, drugs, shots, X-rays, and scans.

I thought a small medical file folder would be adequate for all of us. I soon found out that I needed a large three-ring binder for each of the kids, who up until their teen years had just the usual checkups, childhood scrapes and bruises, shot records, one family doctor, one eye doctor, and one dentist. Best to start off with a large binder and not need most of it than begin as I did and try to organize a sheaf of records after the fact.

Learn where you can find reliable sources of information related to your child's illness, disease, or disability. A friend who had a friend who had the same illness is not always a reliable source. Listen, be open, and find a second reliable source to confirm the information before setting out on a treatment plan based on such friendly advice. Use the library, bookstores, and the Internet. Seek out national organizations that have databases and information regarding the illness or disability. Learn to be a wise advocate and a wise consumer. A healthy dose of skepticism can help you and your child avoid falling for useless or harmful therapies, quack cures, and potions.

The only courage that matters is the kind that gets you from one moment to the next.

—**Mignon McLaughlin**

Four Goals

There is no cure for birth and death save to enjoy the interval.

—**George Santayana, *Soliloquies in England and Later Soliloquies***

In *Soul Medicine: Medical Challenges on Life's Uncertain Journey,* Dr. John E. Postley writes about the four primary goals that must be evaluated for any illness:

1. Cure
2. Extension of the length of life
3. Restoration of function
4. Relief of suffering

These goals might change as the course of an illness evolves. A hoped-for cure might become an attempt to extend the length of life, once a cure is no longer a possibility. Restoring function to a limb might take precedence over pain relief, or vice versa, depending on the diagnosis, treatment plan, and/or prognosis. There are times you might be weighing two or more of these together, with no clear-cut answers, until you actually wade into a murky treatment plan that has a lot of "maybe"s and "if"s involved. *Maybe* a cure will be possible, *if* the tumor is not in a vital area; *maybe* the bone-marrow transplant will work, *if* there are no further complications from the infection; *maybe* the arm will function normally, *if* the nerves are not permanently damaged. *Maybe* she will fully recover from the accident, *if* the trauma to the head is not as severe as first thought.

As you explore the goals with your doctor or medical team, know that it is never wrong to hope for something that is even remotely possible and to share this hope with your child. Keeping muscles supple and developed after a spinal injury is a good idea; finding a way to regenerate nerves in the spinal cord is not beyond the realm of possible science today. Hope, when framed by realistic expectations, can give everyone involved the energy to do demanding, tedious, and sometimes painful exercises, to keep going when it's tough even to get up in the morning.

It is fruitless to hope or wish for something that can never be, that, as Dr. Postley writes, is "beyond the possibilities of the world as we know it." A child born with Down's syndrome will have Down's syndrome all her life; it affects her very genetic makeup. To hope for a cure after the fact is beyond the realm of possibility today. If a parent spends five years searching for a way to solve the reality of the handicap, everyone loses. If the parent can accept the reality and spend that same energy looking for ways to solve the myriad problems that come with that reality, everyone in the family will be better off. Placing blame, feeling guilty, and wishing it weren't true are a waste of precious energy and time. Lobbying the government for funding for special education, contributing to research, finding ways to prevent such damage in other children, working with their families in similar situations, and helping your child become all she can become are constructive uses of both your time and energy.

For us to have wished Maria's tumor was not cancerous would have been fruitless. Whatever it was, it was already. Wishing and hoping after the fact would do no good. What we needed was the strength to accept whatever it was and the energy and resources to do whatever needed to be done. Maria had the optimism that Holocaust survivor Viktor Frankl called "tragic optimism": In the face of the malignancy, she would neither attempt to control this uncontrollable beast nor resign herself to a hopeless fate. She instead accepted the cancer as real, and determined that she could be an active participant in solving the problems that came with it: the surgery, radiation, the scans, the scar, the Synthroid tablets for the rest of her life. In a sense, she surrendered to the reality and thus could use all her energy to get on with her life, knowing that she would be forever changed by her cancer but not controlled by it. She had the courage to speak of her despair, her pain, and her grief, to share her story. Her optimism was not rooted in wishful thinking but grounded in reality. With her optimism and resolve, she was able to do what is known as "reframing": The facts didn't change, her perspective on them did.

She now has a watchfulness and awareness that is not fearful. She is not a cancer victim, nor even a cancer survivor at this point. She is thriving.

For the human spirit is virtually indestructible, and its ability to rise from the ashes remains as long as the body draws breath.

—Dr. Alice Miller, *For Your Own Good*

The Long Good-bye

Death tells us that we must live life now, in the moment—that tomorrow is illusion and never comes. It tells us that it is not the quantity of our days, or hours, or years that matter, but rather the quality of the time spent. Every day is new. Every moment is fresh.

—Leo Buscaglia

Sometimes we must recognize that our goals are no longer a cure, extension of life, or restoration of function. All that is left is to provide relief of suffering as your child is dying. In providing this relief, you can cherish this time with your child and help him find some comfort and peace in his final passage through life. In acknowledging this dying process, you can help your child celebrate his life and the precious gift that it is. One teenager finally said no to any more invasive interventions for his cancer. He was tired of being so sick from the treatments that were not making him well. He chose to live life as fully as he could until he died.

Children who are dying are often the ones to prod their parents along this final passage. They know that their bodies are no longer serving them well. They understand that they are incurably ill and that they are dying. They don't need to hear that everything is going to be okay or that it is foolish to talk about death. They don't need you to change the subject or discount their fears.

When you acknowledge the dying process, your children can share their fears and sorrows openly. They can tell you about their concerns, needs, hopes, and wishes. If there is someone they want to see, you can do your best to make it happen. If you deny the process in an attempt to shield them from the pain, their pain can turn to bitter silence, and you will never know what their last wishes are.

They need to know that they can be angry and express that anger to you. They might be angry at anyone and everyone—angry at themselves for being ill, at their siblings for not being ill, angry at you, the entire medical community, and the world in general. Going back over the three questions for anger can help your child work through the anger and either master it or transform it:

1. Where did it come from? *(Inside myself.)*
2. It is masking another feeling? *(I am hurt, or sad, or frustrated, or disappointed, or afraid.)*
3. Why be angry? *(Because I care. If I didn't care, I wouldn't be angry. I like living, and I don't want to die right now.)*

At times your dying child will probably be deeply saddened. This sadness is not a sadness you want to try to fix. It comes from his grieving his own loss of life. Let him speak his sorrow and be quiet in his sorrow. Your child needs your presence as a sounding board and your touch as a connection to life. Trying to cheer him up and urging him to look on the bright side of things is to deny him his grieving and deny you yours. Anything you feel, know that your child will probably feel it even more. You feel a great loss because you are saying good-bye to him; his loss is greater, since he is saying good-bye to his life.

Allowed to grieve, children can then come to an acceptance of death. Children who have been critically ill for a long time—tired of being weary, tired of being sick—often know this acceptance as a relief that their struggle is almost over.

Your child will begin to separate from those around her, saying her good-byes and wanting to spend more and more time alone or with only one or two people. This part of the passage is often the most difficult one for a parent. You might want her to fight on, to "not go gentle into that good night." Once you can understand that she has not given up, she has accepted that she is near death and is welcoming it, you can spend more of your precious time together providing the simple comforts she needs and being present for one another without all the struggle.

Just like adults, most children are more afraid of the process of dying than they are of death itself. Let your child know that someone will be with her or close at hand at all times as she reaches the final stage of her dying.

As you travel on this painful journey with your child, you will also be preparing her siblings for her death. You will serve as their mentor and guide. There are special issues related to a sibling's death that need to be addressed so that the brothers and sisters of the dying child do not become the "silent mourners" (see Chapter Two: Death).

If I had known what trouble you were bearing I would have been more gentle,
more caring, and tried to give you gladness for a space.

—Mary Carolyn Davies

The Language of Illness

You think you can avoid [pain], but actually you can't. If you do, you just get
sicker, or you feel more pain. But if you can speak it, if you can write it, if you
can paint it, it is very healing.

—Alice Walker, in an interview

Just as there is no easy way to tell your child about a death, there
is no easy way to tell your child about an acute or chronic illness
or a disability. The big difference here is that when you are
telling a child about a death, it is usually *someone else's* death.
When you are telling your child about a disability, it is usually
about *hers.* She feels sick or hurts, or is aware of a problem that
has yet to be given a name. To her, it's about as up-front and as
personal as it gets. No matter the pain, the death was second-
hand; this loss is intimately her own.

The formula is the same: headlines first, facts next, and
delete the editorials. It seems so simple, so matter-of-fact, yet
today as I type this, four years after the fact, I still cry as I cried
the night I sat with my daughter and used that seemingly simple formula: "Maria, you have cancer; the tumor was malignant. You will need the second surgery to remove the rest of
your thyroid."

She knew it might be coming. Nothing I could say could
soften the blow, no introductory phrase would ease the pain.
The easing would have to come later. We cried and hugged, and
then a few questions, a few answers, a few "we'll need to find
out"s, and more "I don't know"s.

After the headlines and facts, be present for your child to hold her and field as many questions as she needs to ask. She might not want to talk right away; she might be too shocked and want only to be alone. She probably won't be interested in much detail, but be prepared in case she is. Assure her that you will get through this together. Diagnosis, treatment, and prognosis can be explained simply as three questions:

Where am I now? (diagnosis)
Where am I going? (prognosis)
And what do I need to do to get there? (treatment)

The language you use with your child will depend on her age and intellectual ability. Keep it simple but truthful. Use the medical terms along with simple explanations or drawings. Children love to use big words, and if a big word like "diabetes" is going to be a part of her life story, the sooner it becomes a part of her vocabulary, the sooner she will be able to connect it with her diagnosis, treatment, and prognosis. Four-year-old Jamie told her preschool teacher that she had "Luke and me," but it was not the same as the Luke in her class.

Language is a powerful tool: It can help your child create a positive, powerful self-image, or it can relegate him to an illness-based identity. As a former teacher, I have seen how easy and efficient it is to shorthand a diagnosis and a child into one entity, complete, with no further explanation needed. A child who has diabetes is *a diabetic;* a child with epilepsy is *an epileptic;* a child with asthma is *an asthmatic.* It takes a bit more effort and a few more syllables to say the former instead of the latter. I think it is worth both more effort and more syllables to keep from defining a child by his illness or disability. Avoid the shorthand and open your eyes to the whole child before you: He might have limits because of his illness, but he is not limited to his illness.

It is much more important to know what sort of patient has a disease than what sort of disease a patient has.

—Sir William Osler

TAO of Illness

Wisdom, compassion and courage. These are the three universal moral qualities.

—Confucius, *The Confucian Analects*

When a child is seriously ill, it can seem as if chaos and confusion reign supreme in the household. The daily routines of the family give way to trips to the hospital, runs to the pharmacy, meetings with specialists whose specialties you can't even pronounce, responding to questions you choke up hearing and choke on answering, scheduling therapies, running to the library and bookstore, searching on the Internet with a browser that doesn't understand your urgency. The insurance company is paying for only a portion of the surgery and none of the experimental therapy. Your sick leave and vacation days are used up. Your child is restless, cranky, and downright miserable. Your other children still need the peanut butter cookies for the soccer team picnic, the dress hemmed for the school play, and your signature on the school field trip form.

The days are too short, the nights too long. You're hard-pressed to give anyone any affection, and your sense of optimism is taking a hit big time. The six critical life messages hung on the refrigerator are hidden under the medical appointment calendar and medicine charts that have taken up permanent residence on the door.

The TAO of Family got you through the everyday events in life, when life seemed so normal. The TAO of Mourning got you through the trauma of a death. Death had a finality to it, its shock mercifully one swift blow. This trauma goes on and on, with new nightmares more vivid than the old. Hopes are raised and dashed; good news simply cloaks the bad news. As much as it seems that these two TAOs have run out the door along with the friends and relatives who don't know what to say or do, they are not gone. You relied on them before when you were at a loss for what to do. They are still there for you to lean on when your mind seems like mush and your heart is full of pain, when your children are scared, fearful, jealous, confused, or embarrassed. The three universal moral qualities—wisdom, compassion, and courage—will join these two to become the TAO of Illness, your road map through what seems to be an endless minefield.

Life only demands from you the strength you possess. Only one feat is possible—not to have run away.

—Dag Hammarskjöld

Ages and Stages

The purpose of life is to matter, to have it make a difference that you have lived at all.

—Leo Rosten

There are so many types of illnesses and disabilities, so many combinations, that it is difficult to generalize about the concerns a child will face at the different ages and stages. And the outcomes are impossible to predict.

The following are the touchpoints or milestones that are characteristic of the normal development of children, intellectually, physically, and emotionally. If your child is five and has a physical handicap, normal intelligence, and normal emotional development, you can use the **five-to-nine** ages and stages as a guideline for what you can expect, what her concerns might be, and what you can do to help her. If your teen has Down's syndrome, he will be physically and sexually developing as a teen and functioning intellectually at the level of a six-year-old. Using the **five-to-nine** *through* the **adolescent** sections as a guide, you can devise your own unique TAO to help your teen through the issues related to his abilities and disabilities. You will need to do the same kind of juggling act if your child has a chronic illness that seriously affects his ability to keep up in school and he is bright enough to know how far behind he is. If you thought at first that you were dealing with a mild handicap, only to have a life-threatening illness thrown in as well, you might need to back up to an age and stage you went through once and address it again.

Infants

Functioning in the present, the infant's primary concern is getting basic needs met now. Infants are agitated when they are held by anxious or agitated adults. They are aware of presence and absence. When ill or in pain, they are unable to express what's bothering them except by being irritable, fitful, cranky, or colicky. They might be lethargic and unresponsive. They often cling to their primary caregiver and resist being held by anyone else, which can place a tremendous strain on the caregiver.

TAO of Illness: Provide loving, consistent care. Respond to your infant's needs with gentle touching, talking, and singing. If you are distraught, surround yourself and your baby with peaceful music and quiet space. Try to stay with usual schedules, rituals, and routines. Incorporate into these any added therapies or medical procedures, rather than having the therapies and procedures take the place of the usual. Be organized, and make treatments as pleasurable as possible.

Use this time to learn as much as you can about your infant's diagnosis, treatment, and prognosis. Knowledge might increase your sorrow. It can also give you the ability to act wisely on behalf of your child. Your attitudes and fears affect your baby's environment. Get the facts, and get support. Your caring and compassionate interactions with your infant will help you come to love her for who she is and who she can become. Learn from your infant to enjoy the now.

Toddlers

Toddlers are actively involved in exploring their bodies, their environment, and their abilities to interact with the world around them. Any illness or disability that gets in the way of this exploration and self-mastery will frustrate them. Any treatment or therapy that restricts them will likely produce a corresponding angry outburst. Getting any toddler to sit still is a difficult task; getting one to sit still for a shot is nearly impossible.

Just as they are able to express a wide range of emotions, toddlers are able to "read" your moods and pick up on your anxieties and fears. They understand more words than they can verbalize. If the illness or disability interferes with the acquisition of language or verbalization, displays of frustration and flares of temper will no doubt be amplified. Sulking, crankiness, and throwing toys are ways toddlers might express this frustration. Toddlers thrive on routine, structure, and play.

TAO of Illness: Just like infants, toddlers need loving and consistent care. To this list you can add play for your toddler and yourself. Play allows both of you a brief respite from the realities of the illness or disability. Therapy can be incorporated into play. Sometimes play is a better therapy than the prescribed therapy if both you and your child need a break from a painful or monotonous treatment.

Your child can now begin to understand his illness or disability. Use picture books, or create your own. Stay away from phrases like "bad cells." Your toddler is likely to equate "bad cells" with "bad behavior" and believe that he did something to cause his illness, or that he deserves it. You can tell him in simple terms what the diagnosis and treatment are, that he is taking medicine or doing a specific therapy to help his body work better. Trips to the doctor will go more smoothly if what will happen is explained in terms that the toddler can understand. Let him take as much responsibility as he can handle: dressing and undressing, putting on a bandage, slipping on the strap for his "funny nose" breathing mask, counting pills, and counting swallows of water. A doll and a play medical kit can help your child act out his treatments and his concerns.

Remember that he understands language better than he can express it. Be aware of what you say in front of him, around him, and when you think he's not listening. Your fears can become your toddler's fears, only his fears are magnified by his inability to put any knowledge and maturity with them. Children who are deaf "read" your body language and get the message even if they don't hear the words.

Don't take your toddler's medical test results as a personal affront to your abilities as a parent. Blaming yourself for poor lab results serves no one. Use your concern to find out possible reasons and possible actions, if any, that can be taken to change the results. Be kind to yourself and to your child. This is not a contest or a fight. Your toddler's ability to live only in the now can help you stay away from the "shoulda, coulda, woulda" trap. Take each day, reflect on it, learn from it, be in it, and then let it go. Tomorrow you can both start again.

Preschoolers

Busy being themselves and getting to know themselves as unique individuals apart from the rest of the world, preschoolers ask lots of questions about their bodies, the world around them, and the people in this world they are exploring. They compare their own bodies with those of others around them, seeking out both differences and similarities. They are astute at picking out the "good guys" and "bad guys" in movies simply by their physical characteristics and tone of speech. Differences are often seen as bad or things to be pitied.

Preschoolers have difficulty separating reality from fantasy. They would rather their imaginary friend take their medicine and take the blame for the broken lamp than take either themselves. They might use "magical thinking"—their belief that their words and actions have a special power—to try to make their illness or disability go away. When you talk about their illness or disability, they might appear sad, bewildered, or ambivalent. They often combine these feelings with their own frustrations and fears and act them out in play or drawings.

TAO of Illness: Take advantage of your preschooler's nonstop questions to explain more fully the headlines and facts of her illness or disability. Use pictures and words to describe the diagnosis and treatment in a matter-of-fact way. Let her know you don't have all the answers, and some questions don't have answers yet. She can count on you to be there for her. Don't tell her that your magic can make the pain of the shot go away. Tell her you know it will hurt and that you will be there for her to squeeze your hand as tightly as she needs to squeeze it. You will be there to hug her and comfort her after it's over.

It's your preschooler's role to use magical and wishful thinking, not yours. Be realistic, hopeful, and responsible in your search for treatments and possible cures. Resist the temptation to hear only what you want to hear or see only what you want to see. If your gut is telling you that your child is not developing or responding the way most other preschoolers are, don't dismiss your gut by denying the reality and making useless excuses: She's shy; she's really able to speak, but she just doesn't want to.

Let her assume as much responsibility for handling the task

related to her illness and disability as she can easily assume. Don't overwhelm her or give her tasks that should rightfully be yours. It's not her job to decide if she does her treatments today; where she does them can be her choice. What color of bandage to use is appropriate for a preschooler; whether or not the bandage is changed is not. Checking off the number of times a day she has swallowed her medicine can be her job; deciding to take the medicine is not. The more responsibilities and decisions she is able to assume regarding the management of her health, the more competent and confident she will feel.

Be careful not to overwhelm her with anything she is not yet ready to handle, be it information about a remotely possible prognosis or choosing her own diet. Listen to her questions and the questions beneath them. Is she concerned about her doll's well-being? Watch her at play to see if she is trying to work through some difficult intellectual or emotional issues.

Aware of her own self-image, it is important that she be able to relate to others of all different ages who share her illness or disability. Play groups and preschool classes can help her develop her physical, social, and language skills. As she begins to go beyond the small circle of family, both you and she will need to deal with stares, questions, and insensitivities of the larger world. Some people will avoid eye contact, and some will speak to you as if your child were not present. Some will express disgust at the uncontrollable noises your daughter is making, while many will be truly caring and at a loss for what to say, fearful of saying the wrong thing. Your assertiveness, wisdom, compassion, and courage can give your child both the support and a model for the response she will be able to use herself someday.

Talk with her about your feelings and show her how she can express her own. Don't overwhelm her with your feelings; she might feel that she can't express her own for fear of hurting you more. Don't stifle your feelings either, or she might never learn to express hers adequately. Don't tell her you know how she feels—you don't. You do know how *you* feel; listen to how *she* feels. There are no "good" feelings or "bad" feelings; there are only real feelings. It's what you both do with those feelings that makes all the difference in the world.

Five- to Nine-Year-Olds

Youngsters from five to nine are into comparing, testing limits, and challenging any and all authority—including that of their doctors, therapists, teachers, and parents. They push limits in hopes that the limits will still be there to be pushed. They need a structure that is both firm and flexible, whether the structure is related to their daily schedule or their weekly therapy. They can separate reality from fantasy. They can analyze, sort, and categorize information and situations. They are aware of the specific elements of their disability or illness that make them different from their peers. If they have a chronic illness, that illness will be a part of their identity.

Frequent absences from school can present special challenges to them. Catching up after three days of school missed because of a cold is easy; catching up while still weak or traumatized after missing two days, then three weeks, then a few more days because of hospitalization is a formidable task. If the disability also affects their ability to learn, the problem with catching up is compounded. And the loss of social contact with their peers during the absences becomes one more card in the deck stacked against their catching up and fitting in.

Youngsters can *begin* to understand the specifics, details, and long-term nature of their illness or disability. Since abstract thinking is still difficult at this age, the facts are more readily understood than their implications. "I need to use crutches to walk" is easy for him to understand; why those crutches affect the way peers approach him is not so easy to understand.

They might challenge you with facts they are learning in school. "If a mother smokes or takes illicit drugs during pregnancy, her baby may be born prematurely" leads to "I was born early, so you must have done something wrong." "If you eat well, your body will grow strong" leads to the accusation that you didn't feed them right when they were little. To them, cause and effect is simple and straightforward. A plus B equals C. "If I am sick and I take my medicine, I will get better" leaves no room for relapses or degenerative diseases. A degenerative disease or relapse is a lousy way to learn that sometimes A plus B equals C and sometimes it equals "back to square one to try something else." The predictabilities they are learning about in math sometimes translate nicely to their illness and sometimes stand only to point out more graphically the unpredictability, unreasonableness, and unfairness of their illness.

They want to fit in with their peers, feel accepted, and develop a few strong friendships. They are able to observe how others feel and often show empathy to children who are "left out" or "singled out," having themselves been in the same boat.

TAO of Illness: The more your youngster is able to be actively involved in decisions related to his treatments, therapies, diet, and medications, the less he will feel controlled by his illness or disability. Taking the initiative to prepare his own snacks for the party at school when the other snacks that will be served will not be good for him not only lets him take charge of his health, it enables him to fit in more easily with his peers. Rather than making a scene or having you take care of these kinds of details, he can begin to take positive steps to make his illness or disability just a part of his life, not the defining characteristic.

You will still need to be the master schedule-maker, but your child can be actively involved in following the schedule posted. Nagging only makes it your problem, not his. Checking the checklists and schedules for therapies or medication at appropriate intervals, or at the end of the day, allows your child to assume more and more responsibility for his own health. Experiencing

minor—not life-threatening—consequences for a missed medication or treatment can be a valuable learning experience.

Children are curious, and they will be less likely to be cruel
to your child if together you and he can do one or more show-
and-tell activities for his classmates. You might want to tell
about a treatment, a shunt, the wheelchair, the breathing appara-
tus, an artificial limb. You could describe a surgery or explain
what a seizure is and how to help. Encouraging your own child
to reach out to help others in need with the skills he has will
reaffirm to him that he, too, can contribute to the well-being of
others and give back some of what has been given to him.

He needs increasingly more details of his own illness or disability. Let him know he can ask any question or express any
concern—that nothing is too difficult for you to hear, nor is
anything too silly to wonder about. "Will I die from this?" can
break your heart, but your child needs to know that if he needs
to ask it, you will hear it. You might not be able to give an
answer, but you will be there to help give him the courage to
deal with his concerns and his fears. "Can my sister catch my
seizures if I sneeze on her accidentally?" can open the door to a
discussion of your child's misperceptions of the cause of
seizures. Or you will find out that Grandma has told him that
his seizures are like when Grandpa sneezes and can't stop—a
well-intentioned explanation, but way off the mark.

Medical language might seem out of reach for your child's
intellectual ability. Use those words, along with a simplified version that matches your child's language ability, and pictures.
Eventually the pictures and the simplified version will be
replaced by the proper medical terms as your child's intellectual
ability develops. The words' having been a part of his heard, if
not vocalized, vocabulary can later help him speak more accurately about his illness or disability. And if he has the accurate
terms, he can search out more information as his need or
curiosity requires. Knowing that he has "cancer" isn't really

helpful. Knowing that he has a cancer called leukemia, which is a cancer of the body's blood-forming tissues, including the bone marrow and lymph system, can help your child grasp the "why?" of a bone-marrow transplant. A five-year-old can say leukemia; a nine-year-old can find the lymph system on the chart in the science book and explain its relationship to bone marrow, when most of his peers are still wondering which bone is where in the body. One eight-year-old asked his doctor, if he exercised more, could he fix his "limp" system. Another asked if eating more starchy foods would help. They were each trying to grasp the language and its meaning for them.

Just as he is shutting the bathroom and bedroom door, demanding his privacy, he will also want some privacy with his illness or disability. The boundaries of privacy regarding his own body seem to be getting violated regularly by doctors, therapists, nurses, and parents poking and prodding, bending and stretching. He is now ready to have input into who can know what about him and his body. His life is not an open book for anyone and everyone to read simply because he has a disability or an illness.

As much as he may not want to let anyone into his circle of confidants, it is important that his teachers, day-care providers, bus drivers, and some adult relatives know the critical elements. But not everyone on the street needs to be privy to the details, regardless of the sometimes prying or insensitive questions or comments. Help him come up with some good assertive one-liners such as these: "It's okay to ask *me*, if you want to know something about me. Mom doesn't always know my answers." "Why would you ask that?" "I don't want to talk about it." "That comment hurt." One of the best responses is a smile followed by silence—a most powerful tool. Teach your child that passive or aggressive comments will not serve him well. Humor has its place as long as it is laughing *with* other people, not *at* them.

When you accompany him to his doctor's appointment or therapy session, let him do the talking as much as he is able. Don't answer questions for him. Don't say, "Sam is concerned about the mobility in his arm," if he is capable of saying it. You can express what your concerns are; he needs to learn to express his. You might want to go over them *before* the appointment and let him jot down some notes to take with him. It's okay to be the note-taker for him in the doctor's office if the two of you agree that would be a good role for you. Let your child lead you on this one. His health will eventually be his full responsibility. Let him practice assuming the role while you are still there to guide and instruct him.

Your child can benefit from seeing people who have the same disability or illness and have overcome many obstacles to lead productive and successful lives. Summer camps that let him enjoy companionship and the outdoors with others who share a similar illness or disability can help energize him physically, mentally, and emotionally to take on the many challenges he faces. He can also help other children who are just embarking on a journey he has been on for some time. The arts, in any form, can help your child express himself and his thoughts, fears, frustrations, disappointments, joys, and sorrows.

Talk to your child about the three passages of grief, how they are not stages to be gotten through but passages that he and you will travel over and over again, not always together or in sync with one another, but you'll both know them by name. A nine-year-old, having spent a difficult year getting his body put back together after a bike accident, hugged his dad and said, "I feel the joy coming back, and it feels so good!" Two years later, after some complications set in, weary from the pain, the boy asked his dad if he thought that he could get that joy back. His dad assured him it was not lost, and that he would find his way there again.

Once your child has the language, not only can he express

what he is feeling, he can see a way through the grief and pain. Reassure him that he indeed must go through it, that he can't go around it.

Preteens

Having spent the last few years collecting information, finding a best buddy, and coming to know *about* their bodies, preteens are trying to use what they learned then to now get to know their bodies, their emotions, and their hormones. Your preteens are not only facing all the emotional, psychological, physical, and intellectual changes that their peers are facing, they will add to this volatile concoction an illness or disability.

The growth hormones and sex hormones might seriously affect the way their bodies utilize the medication they are taking. As well, the medication might seriously affect the growth hormones and sex hormones, sending everything out of whack. What worked when they were children is not working now. And just when they want to be spending time with their friends, they will be spending more time with their doctors, therapists, and counselors trying to get everything into sync, only to have it get out of sync the next week or the next month.

Grappling with issues of their own identity, preteens ask more explicit, pointed, and at times poignant questions about the what, why, and how of their illness or disability. They want the complete details and the long-term effects and implications. Any misinformation you might have given them before to spare them additional pain or grief will be scrutinized. They need to know they can trust you to give them the facts—the good news and the bad news. They also don't want to hear that it could be worse. In the later teen years they will be able to see more objectively that what is theirs to bear is manageable and, yes, they are grateful it is not worse. Right now, to them it *is* the worst.

Body image is very important at this time, and any deviation from the norm can be dreadful. Preteens might deny that their disability affects them at all, refuse to wear their thick glasses and fumble their way down the school corridor, hide the hearing aids in their locker and nod their way through literature class pretending to understand the entire lecture.

Often, the anger, frustration, and hurt they feel is not related to but directed at younger siblings, especially if those siblings have no disability or illness. As much as they might rant and rave at their siblings, they need to realize that just because they have a serious illness or disability, it doesn't mean they aren't responsible for their own behavior. When it comes to their parents, on the other hand, preteens might put their own grief on hold if this is a new illness or disability and they sense that their parents are overwhelmed with their own grief about it. They might expend a lot of energy trying to hold their parents' spirits up, when they really need the reverse: their parents holding them up.

TAO of Illness: Preteens need everything the younger kids need, as well as opportunities to do as many of the normal things kids their age like to do, perhaps with some modifications. Not able to talk with her friends on the phone, your preteen with a severe hearing loss can instead hook up to the buddy program on the Internet and have a two-way e-mail conversation with kids both with and without a hearing disability. Your teen who has a visual handicap can climb the rock wall at the local gym, using her well-tuned hands as her guides; her sighted friends will be trying to figure out how she got to the top so fast while they are still trying to visualize a route to the middle.

The preteen needs to take an even more active role than before in becoming the primary caretaker for her own health. Her medical bracelet is a piece of jewelry she wouldn't think of going out without; granted, it might be partially concealed by her five braided friendship bracelets. You will still accompany her to her doctor's appointment, but she needs time to talk alone with the nurse or the doctor or the therapist.

As you give your preteen explicit information about her disability or illness, be prepared to give her just as much explicit information about her physical development and her sexual development. She might not ask. Don't wait for the questions. Give the answers. And be honest about how her illness might affect her physical and sexual development.

Be just as frank and honest about how being sexually active can seriously affect her health, not only because of her age but also because of her disability or illness. None of us wants to see our preteen or teen pregnant. If your daughter has diabetes and the pregnancy could have serious repercussions for her health and the health of her baby, she needs to know that. Keeping your head in the sand and hoping nothing will happen is foolish, and perhaps life-threatening to your daughter. Waiting until she is "old enough" might be too late.

Just as important is the talk about licit and illicit drugs and alcohol. Alcohol and your preteen's medication can prove to be a deadly mix. Some street drugs and common nonprescription drugs could be toxic to your preteen's liver, already weakened by a serious illness. We know that kids might experiment with various drugs. *Your* kid needs to be well informed about the consequences for *her*. You can tell her that if she's looking for a way to say no without losing face with her peers, then "it will kill me" will probably work.

As she begins to venture away from home more often and for longer periods of time, she needs to be sure that at least one person with her knows about possible problems that can arise because of her disability or illness. What is diabetic shock? What kind of seizures does she have? What foods can't she eat? How can you tell if she is having an allergic reaction? What if she appears disoriented? What to do or not to do? Reassure her that she or her friend can call you anytime either has any questions or suspects that your daughter needs help. If she is experiencing serious pain, discomfort, or disorientation, 911 is the first call, home the second.

You can begin to share with your preteen the concerns you have about her well-being, development, and safety. The life long implications of her disability or illness need to be discussed in relationship to her dreams and aspirations.

Adolescents

Combine the reasoning ability of an adult, the erratic emotional state of an adolescent, and the influence of a major illness or serious disability and you have the makings of a wild roller-coaster ride. The usual ups and downs of the teen years are magnified by the grief, sorrow, and sadness that are the passages your teen is traveling through because of his illness or disability. His girlfriend is now going with his former best friend, his face is full of zits, he had a "puberty check" in English class (soprano on the front end of a word, bass on the back end), his friend from first grade is moving away, and he is no longer on the varsity basketball team. His life seems to be falling apart enough already. Then he gets the lab tests back from the hospital. The white blood count is low again; come in for more tests. Remission is over—he's looking at possible radiation, maybe the bone-marrow transplant as a last resort. Proportion and perspective are lost; everything seems equally awful, equally overwhelming, equally impossible to solve.

A teen who has been living with illness or a disability for some time might react in just the opposite way. Having stared death down and dared it to look him in the face, your teen might be able to bounce back from any adversity, knowing he has survived far worse. He knows death to be the measure against which all of these can be weighed.

As he is trying to break away to become his own person, apart from you and the family as a whole, he recognizes that his disability or illness might make it impossible to break away fully and become an independent adult who can come home just for a visit. The impact of his illness or disability on his sex life, his career choices, even where he can live becomes clearer to him. You've known it, and he knew it in his head. Now he knows it in his heart, and it hurts.

If your teen is severely disabled or severely mentally retarded, he is probably not the one who will reflect on this inability to break away. It will be you who comes full circle with your grief, knowing for real what you couldn't imagine as you celebrated his first birthday years ago: that you will probably for the rest of your life have a dependent child who is counting on you to be there. The reality of old age is not what you dreamed of. There will be no retirement. You might both feel burdened with one another.

TAO of Illness: Teens are risk-takers by nature, but risks that are relatively harmless to others might be serious to teens with a disability or illness. For them, the particular health risks associated with sexual activity, drugs, and smoking need to be made clear. They also need to know that if they get in trouble in any of these areas, their life is far more important to you than a pretense of living up to your expectations, all the while masking serious risk-taking.

Teens who are mentally handicapped are often victims of cruel jokes, comments, and pranks. Their vulnerabilities can be exploited by unkind people. As much as you have taught them to be kind, caring, and accepting, you must also teach them to be wary, skeptical, and willing to be rude if necessary to protect themselves from the unscrupulous among us.

If your teen has an illness or disability that is a part of his life, but not the defining part, you now must take the backseat and allow him to assume full responsibility for his disability or illness—his health, his diet, his checkups, and medical records. You can always offer your wisdom, show your compassion, and demonstrate your courage in the face of your own adversity. You can't live his life for him. He will make his own mistakes, seek his own goals, and reach his own potential. The fruits of his own wisdom, compassion, and courage will be shared with others in his own time and in his own way.

The Journey for Siblings

The journey of the heart for your child who is ill or disabled is not a solo run. His brothers and sisters are along for the ride whether they want to be or not. Not only must they rise to the challenges of growing up, they must also deal with the many challenges they are presented with because they have a sibling who is ill or disabled. The usual sibling rivalry will be there, but it will be amplified, in part because of the amount of parental attention the one child will need for her treatments, trips to the doctor, or extended hospital stays.

Siblings might feel resentment at having to eliminate certain activities, foods, or animals from their home. It's hard to explain to a four-year-old that the cat she has known all of her life has to go live at the neighbors' because her new brother is allergic to cats. Why should a fifteen-year-old not expect his twin to do the same amount of chores he has to do? The fact that his sister has diabetes doesn't convince a ten-year-old that the rest of the family needs to do without chocolate cake for her party. And no explanation is good enough when the family outing is canceled once again because of one more emergency trip to the hospital.

Just as you went through the headlines and facts with your child who is ill or disabled, you will need to do the same with her siblings, keeping in mind their own ages and stages. Be open to their feelings, which might be very different from their brother's or sister's. They might fear that they, too, will get the illness, that they did something to cause it or in anger made a wish that has somehow magically and tragically come true. They might fear that their sibling is going to die soon.

It is difficult to balance your time when one child consumes so much of it. And when siblings feel neglected or passed over, their resentment can build to a point where they will be unwilling to help their sister, or even give affection to her or you. Everyone in the home suffers. They might be angry and resentful that their sister gets so much attention from you. Remember to listen carefully to what they are saying and feeling without placing judgment on their thoughts or feelings. "You shouldn't be jealous of your sister. Do you think she wants to be sick?" will quickly convince a sibling not to express his real feelings to you. "I haven't spent much time with you this last week because your sister was so ill. I've missed our time together, too. Your aunt is going to watch your sister on Saturday, so you and I can have some special time together hiking"—this says you heard what he said and that he, too, is important in your life.

Sometimes it is their own good health that triggers feelings of sadness in siblings. They might feel guilty for being able to climb the mountain, swim the lake, or run the race. It is important that they learn to celebrate their own good health and learn to celebrate the gifts their sister has to offer.

Probably the most difficult challenge for siblings is to overcome the feeling that they need to be responsible for protecting their ill or disabled sister. The mantra "You be the kid, I'll be the parent" can let the siblings know they can be there for their sister but do not need to be the self-appointed monitor of their sister's behavior, health, or antics. The guideline even five-year-olds can understand is this: "If it will only get her in trouble, don't tell. If it will get her out of trouble, tell. If it is both, I need to know."

Teach your children to communicate with one another and to settle their differences themselves, with you as their mediator and guide. Accent the individuality of each of your children. Let them come to know one another as brothers and sisters with their own unique strengths and weaknesses. The wisdom, compassion, and courage possessed by your child with a disability or illness are there for her siblings to possess as well.

The highest wisdom is kindness.

—the Talmud

Part Three

·················

FAMILIES: CHANGES AND CHALLENGES

Chapter 4

Navigating Divorce

There comes a time in some relationships when no matter how sincere the attempt to reconcile the differences or how strong the wish to re-create a part of the past once shared, the struggle becomes so painful that nothing else is felt and the world and all its beauty only add to the discomfort by providing cruel contrast.

—David Viscott

Divorce can be a courageous act or an act of cowardice, or not a choice at all. It can be something that is thrust upon you and you must somehow accept it. Divorce can be good for you, it can be bad for you. It is often ambiguous. But for children, it is a time of great chaos and loss.

Children don't choose a divorce, and they have little control over what happens to them during and after the divorce. They can't control the contact they will have with either parent. The continuity they have come to depend on might be radically altered by a move, by a change in employment of one or both parents, or by financial hardship. The losses are real to children and need to be acknowledged by adults. Kids are resilient, but just because you are elated to be out of a soured marriage, don't expect your children to rally on as if this were just a minor change in their lives.

No two divorces are alike. The impact a divorce has on children has much to do with the maturity, good sense, and goodwill of the parents, as well as the age, emotional and psychological maturity, and gender of the children. In *The Way We Really Are*, Stephanie Koontz explains that divorce in and of itself has a far less negative impact on children's lives than "other frequently co-existing yet analytically separate factors such as poverty, financial loss, school relocation, or a prior history of severe marital conflict." To this list I would add parental conflict *during* the divorce process and parental abandonment *after* the divorce process. It is these factors in accumulation that can have the most profound and devastating effects on children. It is our responsibility as parents, family members, friends, and community to minimize the accumulation of these risk factors for any child experiencing the myriad losses that can surround parental divorce.

We can begin by individually reaching out to offer our presence, concern, and a listening ear; by offering divorce mediation and child advocacy to bring children's rights and needs to the table; and by establishing community programs to help parents and children.

Concerned about the isolation, loneliness, and emotional withdrawal of her students whose parents had died, divorced, or remarried, educator Elizabeth McGonagle created a peer-support group known as Banana Splits to help these children understand and deal with the change in their families. Geared to children ages four to fifteen, the program is used in many communities in the United States and Canada. Another program, Rainbows, developed by Suzy Yehl Marta, a nurse whose three young sons struggled after her divorce, has evolved into programs for children and youth (Spectrum) to help kids deal with the feelings that come up in the wake of a divorce: hope for reconciliation of parents, blame and shame games, feelings of parental abandonment, and anxieties about living in a stepfamily.

Programs like McGonagle's and Marta's help kids "quickly realize they are not the only ones who feel alone and in need of help, and they have the shared experience of their peers." For optimal results, these programs are most successful if both the child and one or both parents participate in concurrent programs of support. And none of the above is a replacement for therapeutic intervention if it is needed.

The Beginning of the End of the Marriage

The act of divorce in itself is not dishonorable; but we are meant to be conscious about the manner in which we conduct ourselves during the process of recanting our vows.

—Caroline Myss, *Anatomy of the Spirit*

The "why?"s of a divorce are not the issue here. Marriages end for many reasons: We were not adequately prepared; we were unrealistic in our expectations; we were madly in love with an emphasis on the "madly," not "love"; the relationship became destructive physically or emotionally; one partner in the marriage fell out of love and into a new love with no effort to address the problems in the first marriage; one partner abandoned ship; both partners realized that the marriage was an impulsive mistake, or both felt that a good divorce was better than a bad marriage. When children are involved, it is important that we rise above our adult differences, act with integrity throughout the divorce process, and make a concerted effort to minimize the accumulation of risk factors that can compound our children's mourning. No small task, to be sure, but an essential one.

You can start your efforts by telling your children as soon as the separation or divorce is a reality, preferably before one parent moves. If possible, both of you can tell the kids at the same time. This will help the two of you keep your story straight and let the kids know that they still have two parents who love them very much. As difficult as the telling is, doing it together might be your last civilized deed as a couple and your first caring act as two separate parents who are committed to coparenting the kids.

Leading with the headlines and the facts works as well here as it does for telling about a death. Give the headline first: "Mom and Dad are getting a divorce." Remember, nothing you say beforehand will ever soften the blow of the headline. Forget lines such as "You kids know that there have been problems in the house" or "Mom and Dad fight all the time" or "Do you remember when the Joneses were fighting a lot, then got a divorce and were much happier?" Kids can hear the headline coming; you do everyone a favor by getting right to it.

We can anticipate the questions and concerns our children will have. Yet just as each divorce is unique, so are our children's responses to the headline. Her husband packing up in the morning to live with his "true love" and unwilling to face the kids, Phyllis struggled all the way to school trying to find the right words to tell her sons that their dad would not be home for dinner. Bracing for the expected tears and wailing, she hugged her boys and told them their dad had moved out, and she started to explain that they were getting a divorce. Eight-year-old Bobby pushed her away and exclaimed, "No, no, he can't do that. He has my soccer uniform in his car! How will I get it in time for the game?"

You might be falling to pieces, overwhelmed by the lifelong ramifications of the divorce. Kids have the wonderful ability to bring you back to the *now*. They are mostly concerned at this point with the immediate effect your divorce will have on them.

After you have given the headline and the facts, you might be tempted to rationalize: "We'll all be a lot better off." "A good divorce is better than a bad marriage." "No more fighting!" "You kids will enjoy both of us more when we are divorced." "This shouldn't be a surprise to you." Don't do it. Just be present for your children to hold them, cry with them, and answer questions they might have. They might be shocked and unable to do anything but cry, or too shocked even to cry. They might want to be left alone, angry at both of you. They probably won't be interested in lots of details at first; just let them know that you will be there for them when they have more questions.

Some of the most common questions are these:

- Where will we [the kids] live?
- Where will Mom live? Where will Dad live?
- Who will keep me safe?
- Will we go to the same school?
- Who gets the dog?
- Will we get to see Grandma and Grandpa?
- Will we be poor?
- Who will take care of me when I'm sick?
- Who will take me to piano lessons?
- When will I see Mom [or Dad]?
- Who will sign my permission slips and my report card?

Older kids may ask the "why?" of the divorce. Save the gory details and dirty laundry for your adult friends, counselors, and therapists. You can be truthful without being hurtful or casting either parent in a negative light. If the question is about the other parent, an agreed-upon response can be "We have agreed not to answer questions on the other parent's behalf. It's okay to ask your dad [mom] about it directly." Remember, the two of you are getting a divorce, not the kids. The kids need to have a meaningful and healthy relationship with both parents. (An exception to this is when one parent is violent, has unremitting addictions, or is seriously mentally ill. See "Peaceful Resolution to a Violent Marriage," page 149.)

The next "why?" is often "Why can't you work it out?" You might or might not have a respectable answer for that one.

However, you can assure your kids that despite the many issues you could not work out in your marriage, one thing you will make work out is a decent coparenting arrangement. The marriage has ended; the family has not. All of you will be bound together as a family—sons and daughters, mom and dad. In the future, the family tree might become a family forest, but even in the forest each branch of this tree will always be connected to its original roots, its family of origin.

If your kids don't ask questions, you can bring things up when you sense they might have an unspoken fear or concern about what is happening. Listen carefully for comments that might contain questions they are afraid to ask. "Dad just up and left. He doesn't care about anybody but himself!" might be a way of hiding "Will he be there for me when I need him?" "How could you just stop loving him?" might be a way to cover up "I'm afraid you will stop loving me, too."

Don't be afraid to be repetitive. What children might have blocked out at the onset of their grief will be listened to and understood at a later time. As children grow older, in an effort to make better sense, emotionally and intellectually, of what they have gone through, they will likely revisit the events and the emotions surrounding the divorce. It's not that they didn't get it the first time; they are now able to think about and process the information at a different level, with a different perspective, and more clearly.

Whether they ask or not, kids need to hear these things over and over again:

- They still have a family.
- They will have two homes, one with Mom and one with Dad.
- Both parents will always love them and take care of them.
- The kids did not cause the divorce. This is an adult problem.

- They will not be left in the dark about any decision that will affect them. Their feelings will be acknowledged and considered. However, the adults will make the decisions, based on the children's best interests.
- They will never be asked to choose one parent over the other, to act as a messenger or as a spy.
- They will not be treated as another piece of property to be fought for, bargained over, or seized.
- They will have the financial support of both parents.

You can let them know that you are truly sorry for the hurt the divorce will cause them, and that you will be there for them to share that hurt. Don't say, "I know how you feel," because you don't. You know how *you* feel. You can show them empathy, not pity or sympathy, by looking at things from their point of view. You can assure them that they have every right to be angry, upset, hurt, and bewildered. They even have the right to feel relieved that the tension they have sensed in the household for so long is finally gone.

Kids' Passages of Mourning

These passages are usually not in sync with the passage either parent is going through. For you, it is the end of a marriage; your children feel as though their family is being ripped apart.

1. Piercing grief of good-bye

"Oh, no. This divorce isn't really happening." During this time kids will do everything they can to deny the reality of the divorce. They will lie to their friends and teachers about what is going on at home. They will try to convince themselves and younger siblings that their parents don't really mean it or are sure to change their minds. They might also express anger and rage at anything and everything. The world is unfair; their teachers are unfair; and their parents are the worst. Either both parents are blamed—"How could they do this to me! I hate them both"—or one parent becomes the villain and one the victim—"How could she do this to my dad? He's never done anything wrong." Schoolwork usually suffers, with kids lost in daydreams, distracted, or so angry and frustrated that they are unable to concentrate on anything but the reality that their world is being torn apart. If parents are doing battle at this point, children might exhibit extreme hostility or aggression.

2. Intense sorrow as they reorganize their lives

"Nothing can make me happy, and I don't want to be happy. What is there to be happy about?" Kids often become lethargic at this point, just barely moving through life. Each major and minor change reminds them every day that the divorce is not going away. It is playing out in 3-D right before their eyes, and they are powerless to stop it.

Some children will try to be very good or overly helpful. Others will try to rewind this horror movie by setting up ways to force their parents to get back together. One girl told her parents they had to watch a film with her as part of a school project—*The Parent Trap*, in which sisters team up to help their divorced parents fall in love again. Some think getting ill or hurt will bring their parents back together.

3. Sorrow that shares space with a quiet joy and gentle peace

Special events, schedule of dates at Mom's house and dates at Dad's house, memories, and photographs still bring pangs of sadness. However, if parents have been able to carry out a decent coparenting plan, have resolved their differences without hostility and vengeance, and have allowed their children to express their feelings and opinions freely and honestly, kids will see that they still have two parents who love them and have their best interests at heart. Their enthusiasm for life will return, and they will be able to redirect their energies into all the normal activities their own growing up entails.

As your children go through their own passages of mourning, don't just speak to them about how they might be feeling, but honor their expressions of those feelings:

"You look really sad."

"It's okay to be angry at both of us."

"We don't have to have the family picture in the living room right now if you are sad when you look at it. Let's put it somewhere else for a while."

"I know you were embarrassed by the questions your friends asked about me and your dad. I wish I had called their parents and the coach and told them myself about the divorce so that you didn't have to be the first to answer their questions. Do you want to talk about what you could possibly say next time? And yes, I will call the coach and your teammates' parents tonight."

"I will tell Grandma and Grandpa about our separation before you visit this weekend and ask them to talk to your mom and me if they have any questions. You don't have to ever play messenger for the adults in this family. You can talk to Grandma and Grandpa about your feelings and what is happening right now if you'd like. There are no secrets we need to keep from them."

When you acknowledge your children's feelings as real and legitimate, without passing judgment, kids learn that their own feelings are important, that they can be trusted to handle them, and that it is okay to count on others for support.

When kids express their feelings irresponsibly, accept the feelings as real, label them, and help your children find alternative expressions that are both responsible and constructive: "It's all right to be frustrated about living at Dad's house three days a week and Mom's four, but throwing your gym bag out the window on the freeway does not serve you well. Talk to me about your frustration, and let's look at how you can solve the problem of the mangled bag and how we can take a new look at the schedule."

How much easier and less helpful to say:

"You must be angry deep down, even though you don't show it."

"Cheer up."

"Don't be so mopey."

"It doesn't help me if you are always sad [angry] all the time."

"The family picture is staying right where it always was. I don't want to hear that it makes you sad."

"Nobody needs to know about our family affairs."

"Just have a good time at your grandparents'. They don't need any more pain in their lives."

"Throwing your bag out the window was stupid. You're just like your dad."

Negating or discounting children's feelings can drive those feelings underground, where they can fester and eventually poison children's outlook on life. Children become so wary of expressing any feelings that none are spontaneously expressed; they must first "check in" with their parents to see if the feeling is okay. They learn to put aside or hide fear, hurt, sorrow, and anger because these emotions get in the way of taking care of the parent's feelings and thus being able to stay connected. They learn not to trust others and to manipulate them for what they need, just as they themselves were manipulated: "If you really are sorry about the divorce, then you will get me the toy I want to make me happy."

Foisting your feelings on them can have similar results. "We are so happy as a family now that she [he] is not around every day, aren't we?" "This holiday won't be any holiday at all; how can we be happy at a time like this?" "Don't cry, I'll buy you a better dog than the one that's at your mom's. Maybe we'll even get two." Letting your feelings and your child's feelings get intertwined, by either smothering his feelings or trying to own the feelings for him, doesn't help him. The child learns to fake feelings, to be confused about what he is really feeling, and to do what he needs to do to please his parents at his own expense. He's like a juggler, trying to juggle one set of feelings at one parent's home and another at the other parent's home, his own feelings left to be dealt with in the darkness of his bedroom at night.

Eventually the child begins to feel angry and resentful at not being truly listened to and cared for. The risk here is that your child will become either too self-sufficient, allowing little if any intimacy into his life, or a needy adult, constantly seeking others to make him feel safe, loved, and secure.

While all of you are going through the passages of mourning, in your own time and in your own way, it is helpful to tell kids you are having some of the same feelings they are having and that it is okay to express those feelings. If you want to give them tools for honoring their own feelings, you can do three things yourself to model those tools:

1. Acknowledge your own feelings and label them: "I'm angry. I'm frustrated. I'm upset. I'm hurt. I'm sad. I'm happy. I'm excited." Note the *"I'm,"* not *"You* make me angry, upset, sad, happy." Own how you feel; no one else can do that for you.

2. Admit that you are angry, frustrated, hurt, or afraid, then do something responsible and purposeful to address these feelings. You don't deny the tears in your eyes; you don't stomp around and then say you're not angry.

3. Reaffirm your belief that you can handle what is happening. "I can do this." "I'll get through this." "I made a mistake, and I will fix it."

Contrary to common belief, divorce does not guarantee that a child will have problems later in life. All children react in some way to divorce, as does anyone whose life is touched by such a major change. But divorce does not foreshadow long-term damage in the way that abuse, neglect, or severe hostility between parents may.

—Dawn Bradley Berry, *The Divorce Sourcebook*

TAO of Divorce

In a world where the family has changed so much and so fast, we need to learn how to live with restraint and kindness and intelligence within the choices we make.

—Suanne Kelman,
All in the Family: A Cultural History of Family Life

As we go through the painful process of the separation and divorce, we need to be present for our children with time, affection, and a sense of optimism. It takes time to teach children to handle their feelings in a way that will serve them well. It takes our affection—that smile, hug, and humor every day—along with our unconditional love. Kids need to know it is all right to feel. It is okay to be happy, concerned, joyful, sad, angry, frustrated, and hurt. Their feelings do count. Feelings are motivators for growth or warning signs that something needs changing. When kids are concerned or happy, they have energy to grow and to reach out to others. When they are angry, hurt, or frustrated, their feelings are signaling their mind and body that something is not right and needs to be changed.

Sometimes what needs changing is not the situation itself but the view of it. This is where optimism comes in. Though they have been brought to their knees with grief, they know they can survive this loss, that people will be there to help them get through their grief, and that they do have the power to choose their own way to respond to it all. With optimism they recognize that they can't always control what is happening to them. What they can control is how they *use* what is happening to them.

Finding the strength to give these things to our children is tough when our own lives are in chaos; we are angry, frustrated, worn out, scared, and confused. We are traveling through our own valley of darkness, and at this point could use some time, affection, and optimism ourselves.

When we are frazzled, it is easy to be lulled into a false sense of comfort and believe that children are resilient and can get through this crisis regardless of what we adults do or don't do. Nothing could be further from the truth. Children grieve in spurts, and so they might at times appear to be unaffected by the divorce. Some might be so concerned about their patents' grief that they will put their own grieving on hold. Children *are* resilient, *and* they need our time, affection, and sense of optimism—as well as the six critical life messages—every day during this time of chaos and loss. Those messages are:

- I believe in you.
- I trust in you.
- I know you can handle this.
- You are listened to.
- You are cared for.
- You are very important to me.

To this TAO of Mourning we must add restraint, kindness, and intelligence:

Restraint: the biting of the tongue, the holding back of the invective and accusations, letting go of the need to get even or win at the other's expense.

Kindness: the benevolence, consideration, generosity, tenderness, and thoughtfulness needed to overcome the bile of bitterness that eats at the lining of the heart.

Intelligence: the discernment, the wisdom to look beyond the vantage point of the earth beneath our own feet and courageously embrace the whole, leaving behind judgmental comments and legalistic righteousness.

When these three are added to the TAO of Mourning, we get the TAO of Divorce.

Helping children cope with separation and divorce is similar to, but not the same as, helping them cope with a death. The death has a finality to it and a concreteness about it; it has rituals surrounding the mourning. There is usually no shame or guilt associated with the death. Although children will often go into denial or extreme fantasies when faced with a death, in the end the death becomes a reality to be accepted. With separation and divorce, things are not so clear-cut.

Sometimes parents are unsure of what is really happening,

themselves fantasizing about a reconciliation. Some parents are in denial that the separation is taking place or feel immense guilt about the circumstances of the divorce and the pain they are causing their children. Others are into the blame/shame game, so busy setting up a battle plan to win the affection and support of their children that the children's needs and feelings are entirely left out of the equation. How you handle the adult issues of the divorce will greatly influence the way your children cope with their loss.

Ages and Stages

Clothe yourself with compassion, kindness, gentleness and patience
—Colossians 3:12

As you go through the TAO of Divorce with children of all ages, remember, the more honestly, clearly, and directly you speak about the divorce, the less likely it is that your children will get stuck in the passages of grief, deny the truth, or blame themselves for the divorce.

Although it is often recommended that you continually remind children that they did not cause the divorce, I would suggest that you spend more of your energy just speaking honestly, clearly, and directly about it, the changes that are happening, and the effects these changes will have on your children. You could easily fall into the trap of seeming to "protest too much," leading children to believe, conversely, that they are somehow responsible for the divorce, since you keep harping on the fact that they are not. It is your time, your affection, and your sense of optimism that are once again the keys to helping your children navigate their own passages of grief.

Unborn child

A child in the womb is aware of its mother's grief, anger, sadness, and depression and might feel distressed, becoming agitated or very still in the womb. Because the two of you are so intimately connected, your physiological responses to grief and mourning affect your entire body and therefore influence the environment of your unborn child.

TAO of Divorce: Provide loving care to yourself, eat as well as possible, and take time to rest. Let others help. For your own sake and that of the unborn child, take time every day to remove yourself from thoughts and activities related to the divorce. In a sense, divorce yourself from what is happening, and take at least an hour to do something good for yourself and your unborn child.

Whether or not you are ready to share the information about the divorce with the rest of the world, it is important that you be frank and honest with your physician. You are divorcing your spouse. Your relationship with your child will last a lifetime. Your unborn child's physical and mental well-being are directly connected, in more ways than one, to yours right now.

Infants

Infants function in the *now*. They are tuned in to their environment and are aware of presence, sudden change in physical or emotional climate, and absence. They might respond to a parent's absence with irritability, crying, fitful or prolonged sleep, and marked changes in eating habits. They might spit up more often or have bowel problems such as diarrhea or constipation. They might cling to their primary caregiver and cry when anyone else tries to take them.

TAO of Divorce: Provide loving, consistent care. Respond to the infant's needs with gentle touching, talking, and singing. Avoid rigid, angry, agitated responses. Avoid loud confrontations or arguments in the presence of the infant. Try to stay with the usual schedules, rituals, and routines. Honor the infant's need for closeness to the primary caregiver; this is not the time to insist on equal time with each parent. This is not a good time to wean your child. The closeness, comfort, security, and familiarity will be welcomed by your child.

Toddlers

Toddlers are actively involved in exploring their bodies, their environment, and their abilities to interact with the world around them. They are tuned in to and can "read" other people's moods. Language is better understood than expressed. They often feel frustrated or angry and express these feelings through temper tantrums, being cranky, sulking, hitting, or throwing toys.

At the time of a divorce, they might become anxious when their primary caregiver leaves for even a short period. They will often regress in both emotional expression and physical abilities, acting more like infants than toddlers. Toddlers are able to understand that one parent lives in one home and the other parent lives in a "new place." They can't, however, comprehend the "why?" of the change. They will often need to gather their possessions around them at bedtime. The security blanket might begin to resemble a hobo pack. They need a safe and predictable environment.

TAO of Divorce: Provide loving, consistent care. Provide limits and boundaries, order and routine. Respond to your toddler's needs with gentle touching, talking, and singing. Avoid rigid, angry, agitated responses. Give your toddler more time to finish eating, dressing, falling asleep, or making the transition from one activity to another. Be tolerant of regression. Don't tolerate abusive or hurtful behavior. Keep changes in daily routines to a minimum in both homes.

Explain the divorce in simple terms, through pictures that you draw, and with storybooks like *Dinosaurs Divorce*. Remind them that they still have both parents who love them very much. Assure them that they did nothing to cause the divorce. Arrange for them to spend time with a relative or adult friend of the same sex as the absent parent. Put a picture of the toddler and the absent parent by the toddler's bed in each home. This might take real resolve on the part of both parents, who will have to view the picture of their ex while putting their toddler to bed. The picture will serve both as a subtle reminder that *you* divorced your ex, your child did not, and that all three of you are in this parenting adventure for the long haul.

Preschoolers

Preschoolers are busy establishing their individual identity, learning new motor and linguistic skills, and figuring out roles and power relationships. They are concrete thinkers, and they equate home with family: To have a family is to have a home. Having a limited concept of space, time, and distance, they have difficulty understanding when Mom or Dad is going to be with them, where Dad/Mom lives, and how they will get back and forth between homes. They are trying to separate fantasy from reality, yet monsters reside under the bed, and Dad will come home if the preschooler "behaves." They feel the loss, experience a wide range of strongly felt emotions, and grieve the fact that their parents don't live together anymore.

Even though they might be very verbal, the ability of preschoolers to verbalize tumultuous feelings will be lagging behind the body's experience and expression of those emotions. Preschoolers might use bullying and aggression, thumb-sucking and sulking to cover fears, anxiety, and sadness. When daily activities or events remind them of the fact that their parents live in separate homes, preschoolers might appear sad, bewildered, withdrawn, depressed, or ambivalent. They might act out those feelings through play, or speak their feelings to an imaginary friend and create fantasy play related to reuniting parents. They might begin bedwetting. They will cling to the primary caregiver, hang on for security, be afraid that this parent will leave, too. The transition from one home to the other can be difficult, and they might express this difficulty by sulking or getting angry just before, during, and shortly after the transition.

TAO of Divorce: Same as for younger children, plus talk about concerns they might have related to the divorce. Explain the changes that are happening. Be available to answer the same questions over and over again. Share books like *Dinosaurs Divorce*. Preschoolers will often transfer the expressions from the book to their own life experiences and use the phrases to explain what is going on in their own heads and hearts.

Because of the preschooler's belief in "magical thinking," you need to reassure her that nothing she said or did—or didn't say or didn't do—caused Mom or Dad to leave, and nothing she says or does will make Dad and Mom get back together. It is an adult problem, with adult answers. Take time every day to let your child know that both parents love her very much and are willing to listen. Provide a peaceful and gradual transition as she moves between Mom's house and Dad's house.

Five- to Nine-Year-Olds

Even at the best of times, youngsters from five to nine invest a lot of energy in challenging parents' values, testing boundaries and rules, and arguing with "authority." When they learn of the divorce, their anger may exacerbate this behavior. They are often extremely angry with both parents for divorcing. They might take out that anger on whichever parent they are with, blaming him or her for making their life miserable. Yet they can still be openly affectionate and might show concern about their parents' well-being.

At this age, children are beginning to understand the many implications of divorce. Though they might not want to believe it, they realize that their parents don't love each other and are not getting back together. They have fears about abandonment, money, where they will live, what will happen in the future, and they are able to express these fears verbally. Feeling vulnerable, they might deny that the divorce hurt them in any way; they will act tough to hide the pain. They will play with great passion, trying to drive the pain away, or they might have intentional "accidents," either to punish themselves for causing the divorce or to get both parents' attention and affection.

Feeling self-conscious and embarrassed about the divorce, they might create elaborate stories to explain away the fact that their parents are living apart. They might feel a need to take sides in what they perceive divorce to be: a fight. They might also have a significant decline in school performance during the first year after the separation or divorce.

TAO of Divorce: Kids at this age will often ask many specific questions about what caused the divorce, what is happening now, and what they can expect in the future. They need to know that you believe that each and every one of you will make it through this crisis. Remind them that finding fault is not the issue in the divorce, and that both parents love them very much. If they are using their intellect to cover their feelings, they need to be encouraged to express and at times vent their feelings about the divorce with you, another relative, or another trusted adult.

Talk with them about the passages of grief and what they can expect in those passages. Participating in peer-support programs such as Banana Splits can help them see that they are not alone in this kind of family change. Youngsters in such programs help themselves as they help others who are just beginning their journey through the changes. Drawing, painting, or molding clay can help them express feelings that they find difficult or are afraid to express. Be cautious not to overwhelm them with all of your anger, hurt, and grief. Find an adult to confide in about your adult issues, your frustration, financial concerns, and your present perceptions of your ex-spouse.

Preteens

Unable to be adults, unable to be children, and wanting to be both, preteens are shier about crying or seeking out hugs. They yearn for all the independence that comes with being a teen, to be left alone to figure out how they can deal with the tumultuous feelings they have.

Busy with all the new academic and social activities that are the benchmarks of this age, the preteen might direct a lot of anger, frustration, and hurt at the parent she believes is responsible for "destroying the family stability and security" she had come to depend on. Preteens can understand but rarely accept the divorce. They might attempt to take the place of the absent parent or begin taking care of the parent they are living with, helping Mom or Dad through the grief passages and putting their own grief on the back shelf, to be taken down when the parent is strong enough and not so needy anymore. They might vilify the parent they feel is responsible for the divorce and create a strong alliance with the other parent.

Preteens want the facts about how the divorce will affect their everyday life. At the same time, they will deny that it affects them at all. They might sulk, say they don't feel anything, say they don't care, in an attempt to bury their intense pain and their unsettling fears about the changes that are occurring in every area of their lives. They might make exaggerated attempts to help others while being superstrong themselves, as though this display of strength will keep them from being tainted by the divorce.

They fear being different from their peers and might have a sense of shame concerning the divorce. They might complain about headaches, stomachaches, inability to sleep. They might be embarrassed or angry about changes (or perceived changes) in their parents' sexual behavior.

TAO of Divorce: Preteens need everything the younger kids need, as well as opportunities to spend time with their peers to talk, to laugh, and to have fun. Since they need their parents to be their mentors, it is important that you express your feelings with them and show them how you are doing everything you can at this point to get things back on track. They need your reassurance that family life will go on, albeit in a new form, and that all of you will make it through this trying time.

It is tempting to criticize preteens when they revert to younger behavior or act silly. Be tolerant of regression. Don't tolerate abusive or hurtful behavior. In your optimism, you have to believe they are doing the best they can right now.

Confront attempts to idealize or demonize the absent parent, but do it without either derogatory comments or platitudes, reminding them that we all have strong and weak points, assets and liabilities. Let them know you care about their input concerning possible changes that might have to be made. Assure them that you will listen and take into account their feelings and ideas, even though, in the end, it will be the adults who make the final decisions that they believe are in the best interests of the preteen. Don't ask preteens to establish the visitation schedule, decide on a new home, or pick a new school. Do let them help make each of their homes a place they can feel they belong, where they are comfortable and safe. Be open to talking about their feelings of shame and embarrassment. Be alert to risk-taking and rebellion that is persistent or extreme.

Adolescents

Take all the possible responses of the preteen and magnify them with the reasoning ability of an adult and the erratic emotional states of a teenager, and you have the common characteristics of adolescents confronting the reality of their parents' divorce. Just as preteens are in transition from being children to being teens, adolescents are on the cusp of being adults. Emerging as separate, independent persons with their own identity and values, they are able to understand divorce but would rather not accept it at this crazy time in their own lives. They might seem to care less about the divorce because they are so caught up in their own relationships outside the family. They might deny that the divorce even bothers them at all.

Younger teens tend to idealize their parents and the marriage that is ending. Older ones might already be experiencing conflict with their parents, and the divorce might make the conflict more intense. Emotions are in turmoil; moods change abruptly and swing erratically from sadness to anger, giddiness to anguish. Teens might try to exploit both parents by making unreasonable demands. They often share their thoughts and feelings with close friends and clam up when around adults. They might try to be supergood or superbad. Their normal search for independence might leave them less family-focused and more inclined to want to be with their friends. Some retreat into mind-numbing depression or use drugs, alcohol, or food to drown the pain. They might join a gang or religious cult in order to gain a sense of belonging. They might get into trouble with the law. They might be anxious and cynical about intimate relationships, rail against the inequities and unfairness in every area of their lives, and swear never to get close to or marry anyone themselves.

Adolescents often feel caught in the middle or feel a need to be loyal to the parent "suffering" the most. They might be afraid to separate from one or both parents, feeling a need to be there to "shore them up." They might be overwhelmed with the added responsibilities for younger siblings, running the household, and the need they feel to "parent" their parents.

They might seek to escape the trauma of their home by setting up their own home with a boyfriend or girlfriend, chaotically trying to create a new, stable family.

TAO of Divorce: Adolescents need everything that preteens need from you. As well, they need you to be there as their mentor, modeling restraint, kindness, and intelligence. They are not your confidant or best friend.

Adolescents are very aware of the dynamics of the family and need you to be honest and forthright about what is happening. Encourage them to express their feelings. Be alert to signs of deep depression, severe fatigue, alcohol or drug use, overeating or undereating, movement toward gangs or religious cults. As they are striving to become independent, they need to know that even in this time of chaos and loss they can count on you for time, affection, and a sense of optimism.

The Language of Family Versus the Language of Dysfunction

Language plays a fundamental part in shaping our reality, our values, and our priorities, in giving substance to our beliefs, defining our status and our roles. It also defines the values of our culture. What we call ourselves and what we are called can shape what we become. Words exist to serve, not to oppress.

—**Isolina Ricci,** *Mom's House, Dad's House*

Words are powerful tools, and the language of divorce needs to be cleaned up if we are truly to help our children develop a strong sense of self. Without even thinking about the connotations of words and their impact on children, we say that they are from a "broken home" and the "victims of divorce." The phrase "she's from a single-parent home" is often used, not to describe the makeup of the family system but rather to explain, excuse, or validate a flaw in the child's behavior or performance.

In her book *Mom's House, Dad's House*, Isolina Ricci attacks our cultural attitudes toward families who have experienced a divorce. She demonstrates how the words we use to express this attitude tend to isolate parents and their children, brand them, undermine their sense of self, family, and parenthood, and connote a second-class status worthy of pity or condescension. Adversarial language that is harmful and degrading perpetuates the suspicions our culture has about family systems that are different from the "nuclear family" or "family of origin." Ricci suggests that simply by changing certain everyday words we can improve our own and our children's sense of identity, sense of family, and sense of belonging in a community:

Old Language	New Language
• broken home	• my family, two-home family
• failed marriage	• marriage ended
• custodial parent/ noncustodial parent	• parent, mom, dad
• joint custody	• coparenting, shared parenting, shared responsibility
• sole custody	• primary responsibility
• children are visiting	• the children are with their dad, with their other family
• custody and visitation	• live with, be with, stay with

Old Language

- access
- ex-wife, ex-husband

- real family

- child support

New Language

- parenting time
- children's mother or children's father; helps focus on the parent/ parent relationship you will continue to have, not on your former intimate relationship
- family of origin, nuclear family
- contribution

Simply changing the language does not eliminate the hurt and the pain all of us must deal with when there is a major disruption in our lives. Changing the language doesn't negate the fact that a divorce has taken place; it does not deny the realities of the temporary loss of structure and stability and perhaps status that existed in the marriage. It does not deny the chaos and confusion our children faced. We've experienced a great loss. But once we've mourned that loss, it will not serve us or our children to constantly define ourselves only in relationship to that loss.

What we call ourselves and what we are called can greatly influence who we become. Using the term "victim of divorce" can keep us feeling victimized, helpless, and despairing. Just using those words over and over again can lead us to believe that our future can be seen only in reference to the divorce. We become passive victims open to further victimization. And now we have a great excuse for any irresponsible acts we might commit in the future: "My parents divorced when I was three; my life is a wreck because of it. I only drink to drown the pain."

Calling ourselves "survivors of divorce" forces us still to see all future events filtered through the distorted lens of the "divorce event." Although survivorship is a step beyond victimhood, both bind us up forever in the divorce. They do not let us go beyond it.

A change in language can help us move on to a life that acknowledges the divorce as a part of our past but refuses to let it put a stranglehold on the present or the future. Family therapist Constance Ahrons defines the family constellation created as a result of divorce as a "binuclear family"—that is, a family that spans two households. Whatever you as a family choose to call your constellation, make it a positive and realistic definition that is meaningful to all of you.

Stephanie, a ten-year-old who was living with her grandparents after her adoptive mother was diagnosed with cancer and her adoptive father had abandoned the family, took home six worksheets for her school's family-tree assignment. Together with her grandmother, she colored each tree a different color, pasted all the sheets together, and happily presented "Stephanie's Family Forest" to her class. Other children in the class asked if they, too, could take home extra worksheets to portray their families accurately. One boy returned with a colorful array of trees to which he had added all of the animals in his different homes. He had neatly labeled his project "My Family Jungle."

Legal labels can also make a difference in the tone that is set for how we will coparent. "Having custody" and "visitation" tends to set up a winner/loser scenario. To say that one parent won custody and the other has visitation rights tends to place more value on one parent's contribution than the other's. No one wants to spend the next twelve years just "visiting" his or her kids. An evenhanded description—"Billy and Samantha will live with their father every Thursday beginning at 5:00 P.M. through to the following Monday at school time. The remainder of the time they will live with their mother. During the summer they will live with their father"—helps all parties involved in the parenting agreement to see that both parents play a vital role in the parenting of the kids. Joint and shared responsibility is a more accurate description than joint custody. It does not necessarily mean a fifty-fifty split. It means that both parents recognize that each parent's contribution is important and both are willing to work together to make each one's individual contribution meaningful and functional.

It is not only societal and legal language that can be destructive. The words we speak in our own homes about one another can be equally destructive, as well as scary, to children. These words serve no useful purpose and have no constructive alternative:

She was no good.

He's a lazy bum.

You're just like your mother.

It's a whole lot better with him out of our lives.

You better toe the mark, or I'll send you to live with your father.

What did he say about me?

If he loved you, he wouldn't have left.

If she loved you, she would send the child-support check on time.

If it weren't for three kids, I could be as carefree as your father.

There's more than one fish in the sea.

I wish he was dead; it would be a lot easier for all of us.

Men are no good.

You can't trust women.

Don't get married. Do something better with your life.

If your dad is late again, I'll go to court and make sure he loses his visitation rights.

Who do you really want to be with?

I get the hard work of raising you kids; all he does is have fun with you.

If you want a bike, ask your father. He has money to burn.

All of these are best left unsaid. If you have a great desire to say them, write them down, reread them, then tear them up or burn them. Or hop into the shower, spew out the invective, and listen to your words amplify and bounce back at you. Better yet, get at the issues and feelings behind your words and figure out a way to handle them constructively and responsibly.

Do not let any unwholesome talk come out of your mouths, but only what is helpful for building others up according to their needs, that it may benefit those who listen.

—Ephesians 4:29

Options for Constructive Solutions to Divorce Disputes

Willing good occurs when, upon due reflection, we will ourselves to do and speak that which we know to be right, even when the burden is heavy; willing evil occurs when we do anything else. . . . Evil is not simply the result of a decision to do a bad thing; it is refusing to make a decision to do a good thing.

—Stephen L. Carter, *Integrity*

Regardless of how you and your marriage partner related or didn't relate in the marriage, during the process of divorce you will need to talk to one another, make decisions, put piles and piles of information together, and discuss the next twenty years of your children's lives—all of this when neither of you feel particularly civil, friendly, talkative, organized, clearheaded, or receptive to the other's presence. In his book *Integrity,* Stephen L. Carter sets out three steps that can be useful in helping the two of you do what is right and just, not necessarily what you are legally entitled to do, obligated to do, or want to do:

1. **Discerning what is right and what is wrong.** This step requires a degree of moral reflectiveness, a rising above what you might think or feel you want. It demands of you mindfulness and a wise heart. Withholding support payments to get back at your kids' other parent for marrying your best friend might be what you feel like doing, but after taking time to discern what is right and what is wrong, you send the check.

2. **Acting on what you have discerned, even at personal cost.** This step involves making and keeping commitments. Continuing your monetary contributions for a disheveled, disgruntled, mouthy teen because you made a commitment to do so when he was a happy-go-lucky five-year-old might not feel good, but it is a commitment you made, and keeping that commitment to your child is the right thing to do.

3. **Saying openly that you are acting on your**

understanding of right and wrong. This step is a reminder that people of integrity are unashamed of doing what it is they have discerned is the right thing to do. Suffering comments from your friends, family, and new spouse about how foolish you are to send the money, you do not make an attempt to sneak the check past anyone who might deride you for such foolishness.

These three steps will not provide clear-cut answers for you. They are like the TAO of Family: They can provide a formula for the both of you, even with your anger and hurt, to work together with integrity to resolve the complex issues facing you and your children in divorce.

You have three legal tools available to you. They are, in descending order of civility and ascending order of cost, mediation, arbitration, and litigation. If used exclusively to resolve all of the issues, the first will enable you to establish a workable coparenting arrangement but might not adequately address complex property and financial issues; the last used alone will guarantee that you get what you are legally "entitled to" and that both of you will be working for years to pay your legal bills—and even longer to repair the emotional rifts created by the cycle of retribution and retaliation that litigation can invite.

It's possible that you will need to use all three of these options as you sort out your different personal, material, and financial arrangements. In order to discern which option to use when, you will need first to separate the major decisions about your kids from decisions that need to be made regarding money and property. Attend to kid issues first, and let these issues be the overriding factor in any decision you subsequently make about property and money. In the heat of battle, it is easy to want to do just the reverse, try to settle the property and money issues and then *use* the kids as clubs to beat the other parent into giving in or giving up.

Once the priority is set, you will need to define personal and material issues and discern how they relate to one another. Equal time in both Mom's and Dad's houses might be important for a seven-year-old, while a stable home and consistent caregiver for an infant will override the desire both parents have to spend equal time with the child. This is the time to let go of the need to "win" on principles and work toward what is in the best interest of the kids. Compromises and negotiations will become a way of life for all of you.

Mediation

Mediation is used where there is a likelihood that the parents will be able to reach an agreement with the assistance of a neutral party. It is a voluntary settlement process; it does not favor one parent over the other. Its goal is to move both parties to reconciliation and agreement. It is the most effective of the three options when parents are willing and able to put aside their emotional differences, anger, and frustration to work toward developing a coparenting plan.

It does not replace or supplant the protection that the legal system can provide. In fact, it works within the legal system. Any agreements reached in mediation are reviewed by each parent's attorney and included in the legal divorce decree.

Mediation does not have to be used for every issue; some issues might need to be arbitrated or litigated. It is the best of the three options to use when the parties involved will have an ongoing relationship and the agreements reached will affect that relationship. Parenting, being a lifelong commitment, does require an ongoing relationship. Once the pension plan is divided, the boat sold, and the dishes stored for the kids to hassle over at a later date, neither party need deal with those issues again. But you will both want to be at the parent conferences, sports events, graduation ceremonies, and weddings of your kids. It helps if you are on speaking terms with one another.

Mediation can increase that likelihood. Mediation can save both of you time, energy, grief, stress, lost sleep, angry words, heated exchanges, and money. It keeps as its focal point the best interests of the kids. It promotes a genuine peace.

Mediators do not give legal advice and do not make decisions for either parent. They help parents develop their own agreements. They can help to ease tensions and create an environment where divorcing partners can begin to practice coparenting communication skills. They help both parents practice effective listening skills, develop a willingness to be patient with one another, compromise, and be open to change. Having these skills does not ensure that coparenting will proceed without a hitch. It is a process, not an event, so it will have its turbulent times and times of genuine peace and calm. Just having the tools and confidence to use them can help give both parents courage to hang in there in the rough times for their kids' sake.

Is there ever a time when mediation is not a constructive option? Yes. If you are leaving a violent or abusive marriage, attempts at mediation could put you and your children at greater risk of emotional harm or physical abuse. (See "Peaceful Resolution to a Violent Marriage," page 149.)

Arbitration

The basic difference between arbitration and mediation is that arbitration involves a decision by the arbitrator, while mediation does not. In mediation, the mediator helps the two parties arrive at a mutual agreement. In arbitration, the arbitrator listens to both sides and then makes a binding decision based on the arbitrator's understanding of the dispute. Arbitration works best when there is no reasonable likelihood of a mutually negotiated agreement and the parties in the dispute will not have a continuing relationship related to the issues in dispute. When the boat is sold and the proceeds split, either or both of you can go out and buy a new boat, never set foot on a boat again, or use your half of the money to take flying lessons. Neither of you has to return to the issue of the boat again; it's over, and you are moving on. This is far more difficult with your coparenting agreement—it goes on and on.

If you have tried mediation and it was unsuccessful, arbitration might help both of you, during a time of great emotional upheaval, to begin to see the conflicts you are having with one another in a more objective light. The arbitrator will have the best interests of your children in mind and at heart when he or she makes the decisions you were unable to reach together.

Many parents find that they are able, a year later, to return to a mediator and successfully mediate necessary revisions to a coparenting agreement. The anger and the hurt that kept them from even being civil toward one another have given way to a mutual concern for the well-being of their kids. They begin to see one another as allies, not obstacles.

Litigation

The adversarial process is designed to resolve legal issues, not reconcile people or help them change the attitudes and habits that led to the dissolution of the marriage. It might be necessary to use this process to resolve material disputes about property, pension plans, Aunt Hilda's china, or the family boat. It can be used judiciously to make sure both parties receive a fair share of the assets of the marriage. However, not only does unleashing the full force of the legal system during the divorce allow both of you to contribute heftily to your favorite lawyer's retirement fund and summer getaway in the Greek Isles, but the ensuing battle will be certain to do six less than desirable things that will undermine your ability to establish a functioning coparenting arrangement:

1. **Create a victim/victor mentality.** ("I won custody of the kids." "We have to divide all the spoils of the marriage.") The conflict is viewed as a fight in which there is a winner and a loser. If and when the fight is over, nobody really wins, and the kids end up losing the most. Kids often feel responsible for causing the battle and assume guilt for the outcome. Or they get tired of being pawns in the battle and distance themselves emotionally and/or physically from both parents.

2. **Distort your perceptions and allow you to fanaticize one another.** ("She was a horrible mother." "He was a lazy, no-good bum." "She's an unfit parent." "He's a drunk.") As your perspective becomes more distorted, you are apt to imagine the worst in one another. When you imagine the worst in one another, you are more likely to misjudge each other's motives and deeds. ("He only paid the support because he knew I was taking him to court." "She didn't *forget* to tell me about the school play, she didn't want me to see my kid perform.") When you begin to see everything the other says or does from a negative perspective, you become bitter, resentful, and antagonistic. This attitude starts to color all of your relationships, including your relationship with your kids.

3. **Oversimplify complex issues.** ("Everything will be split down the middle—I will take one of the twins and you will get the other. The only fair split is an equal split." "If you move away, you won't ever see the kids again." "If you bring a man into the house, I'll take the kids away." "No pay, no see." "This is the schedule, and it cannot be changed. I don't care if it is a family reunion; it's my week, not yours.") Material issues and personal issues become enmeshed. Issues are viewed as black and white, right and wrong, with no thought

given to the complexities of relationships. Creative options for resolving conflict are not explored.

4. **Distort the truth.** ("He *never* paid the bills on time." "She was at least two hours late picking the kids up. I had three hysterical kids on my hands." "The kids spend the whole weekend partying with their dad and that woman.") Reality: two days late with the bills three times; forty-five minutes late, kids were sulking; birthday party after a soccer game. The distortions of truth are often veiled attempts to diminish the reputation of the other parent, and they serve no constructive purpose.

5. **Create a cycle of retribution and retaliation.** ("I'll make her pay dearly for this divorce." "I'll take the kids so far away, she'll wish she'd never challenged me in court." "He thinks he won that round. I've got enough damaging stuff on him so his kids will hate him forever.") Trying to hurt one another through the kids can only backfire on both of you. As well, it puts the kids in the painful position of trying to be loyal to both of you.

6. **Consume an exorbitant amount of time and energy at a time when both are at a premium.** A cartoon showing Moses coming down the mountain with a ton of tablets has as its caption "This time I met lawyers." Litigation involves a lot of paperwork that takes a lot of time and energy to complete. A lot of time is spent waiting to get a court date to have your side heard, and often much of the time waiting is spent developing a case against the other parent. Such negative use of time and energy virtually guarantees that time, affection, and a sense of optimism for yourself and your kids will go out the window, and legal bills will continue to mount.

Litigation often takes a much higher personal, emotional, and financial toll than most people ever imagine. You might win, but the cost, in the end, will probably far outweigh any gains. I can promise you, the toll on your kids will be bigger than any judgment in your favor.

My mind will not experience peace if it fosters painful thoughts of hatred.
—Shantideva, *A Guide to the Bodhisattva's Way of Life*

Coparenting Plan

Coparenting depends on setting up new emotional boundaries and allowing your children to have their own emotions, identity, and choices. It requires leaving the past in the past, and focusing on the present and the future. Most importantly, it requires never forgetting the vision that you are working together for your children's greatest benefit.
—Elizabeth Hickey and Elizabeth Dalton, *Healing Hearts*

The same philosophical tenets of parenting explored in *kids are worth it! Giving your child the gift of inner discipline* can become your philosophical tenets of coparenting:

1. Our kids are worth the time, energy, and resources it takes to come up with a coparenting plan that demonstrates both parents' commitment to do the right thing for the kids. We have a clear sense of direction and purpose.
2. We will not treat the kids or their parent in a way we ourselves would not want to be treated. We will be honest and fair.
3. Whatever we do will leave the dignity of all parties intact. We are willing to compromise, to agree to disagree, and to confront one another assertively.

It is possible to end your marriage and create responsible, civilized, active two-household families if you are willing to reduce the acrimony between one another and remain involved in your children's lives. Not every divorcing couple has the trust, respect, and ability to communicate with each other to make a coparenting arrangement work. But for the sake of the kids, it is worth investing your time, energy, and patience to try to create a coparenting plan.

The process itself can help the two of you remain focused on your goal of minimizing the accumulation of risk factors that can have devastating consequences for your kids and maximizing the opportunities to allow your kids to develop a strong, loving relationship with both parents. In the beginning, you might not actually be coparenting. Parallel-parenting might be more like it, as each of you tries to disengage from the other and begin a new chapter in your own life, apart but forever connected as parents by and to your children.

There are many legal and personal issues involved in developing a long-term coparenting plan. A mediator can help the two of you create a plan to meet your individual family's need. But a mediator can only help. You would be wise to reflect, in advance and often, on the following issues:

- Basic philosophical tenets that each parent espouses.
- Your individual and joint parenting goals.
- Responsibilities each of you are going to assume.
- Decisions each of you are going to be expected to make.
- Responsibilities the children are going to be given, and by whom.
- Decisions the children can make.
- Resolution of present and future disagreements. (It is helpful to have a Plan A and a Plan B for this one.)
- The risk factors to the kids and ways each of you plans to minimize them.

- The outer limits of tolerance in the coparenting plan, i.e., what each of you can live with, can't live with, could tolerate, don't mind, and don't you even ask. (You'll be amazed how this one changes over time.)
- Money.

All of these issues can be addressed simply, but by no means easily, using the six questions you studied in grammar school: Who? What? Where? When? How? and Why? Coming up with some tentative answers will help the two of you come to the mediation with a greater understanding of the intricacies and possible pitfalls involved in moving from a two-parent home to a shared parenting commitment.

If you are exhausted after only reading the list of issues that need to be addressed in a coparenting plan, know that you will be even more exhausted creating the plan with the person you no longer want to spend the rest of your life with, and even more drained at times trying to carry out your part of the plan.

The two of you might never have discussed many of these issues and somehow managed to handle them before the divorce. Special days, special occasions, holidays, visiting grandparents, signing report cards, taking the kids to the doctor's, paying the bills, finding child care, and creating family traditions came about with very little conscious thought. Now even a telephone schedule has to be established, and time-sharing arrangements must be written down. As well, you get the opportunity to grit your teeth as your youngest returns from the other parent's home proudly modeling the new cowboy pants and shirt the other parent's new lover has bought for him.

It is time to revisit the TAO of Divorce: time, affection, a sense of optimism, the six critical life messages, along with restraint, kindness, and intelligence. Practice what you have been giving your children. Remember to care for yourself. The greatest thing you can do for your children is to get your own act together.

The best upbringing that children can receive is to observe their parents taking excellent care of themselves—mind, body, spirit.

—**Dr. Benjamin Spock**

Assertive Confrontation

Speak when you are angry and you will make the best speech you will ever regret.

—**Ambrose Bierce**

People who fight fire with fire usually end up with ashes.

—**Abigail Van Buren, syndicated columnist**

In anger, hurt, or frustration, your children might say ugly things about their other parent. You might be inclined to reinforce or elaborate on their comments. You might even be inclined to cheer their outbursts. "I'm glad you see what a jerk your father is." Don't. And don't try to defend the other parent or make excuses. "Your mom probably didn't mean to forget to pick you up. She is trying to juggle so much right now." Let kids vent their feelings. You can help them learn to speak what they are feeling without attacking another person. "It is all right to be angry about what your dad did this weekend. It's not all right to call him ugly names. Let's talk about a way you can express what you need to say without ripping your dad apart."

The issue might be so explosive or divisive that it is necessary to use the tools of assertive confrontation. If you have practiced these yourself, it will be easy to teach your children. If you have not, use this as an opportunity to get familiar with the process so that you can take the high road when it comes to your conflict with your kids' other parent. It helps kids greatly if you serve as a model, showing them a way of confronting someone else that demonstrates that you accept your own feelings and have a responsible way of addressing them.

In order to confront a parent, kids need to understand their anger by asking and answering three questions:

- Where did it come from? *(From inside myself.)*
- Is it masking another feeling? *(I am hurt, or frustrated, or disappointed, or afraid.)*
- Why be angry anyway? *(Because I care. If I didn't care, I wouldn't be angry. I can't be angry about something I don't care about, with someone I don't care about.)*

Once the anger is understood, kids can use the seven steps for a fair fight to construct a productive, assertive confrontation:

1. When you are upset or angry, say so in an upset or angry tone of voice. Let your whole body speak the message in a straightforward, assertive manner—not aggressively or passively. Say what you are feeling in a firm voice. Your tone can convey anger without your voice becoming a loud scream or a shaky whisper.
2. Tell the other person about your feelings. "Dad, I am angry about what happened last weekend." It is critical that your body and mouth are saying the same thing. To slam the door in his face and say that you are not upset about anything is passive-aggressive. To say in a soft whisper that you are angry is not to involve your body in delivering the message. In fact, if you quietly say you are angry, anger is probably masking another feeling, such as disappointment or sadness. You then have to ask if your body is speaking your true feeling.

3. State your belief out loud and avoid killer statements. "I believe it is important that you keep your promises about making time for me to do something with you." Avoid comments such as "You never keep your promises." "You care more about your woman friend than you do about me." "You are a jerk." These killer statements only attack the other person. Stick to your belief. This step is often left out of confrontations, yet it is critical that the person whom you are confronting know what your belief is.

4. Close the time gap between the hurt and the expression of the hurt. Give direct feedback. "You said we would go to a ball game this weekend. Instead I had to spend time with you and your friend shopping for your dinner party." Tell the other person what he has done. Don't tell him he never keeps his promises or that your mom always says what she means and means what she says. Not only do these put your dad down, they are likely to arouse resentment and anger. Don't tell him he also forgot to pick you up three months ago and missed your recital two months ago. To tell him that you are still angry about several other things he did a while ago is both unproductive and an indication that you need to work on dealing effectively with your anger.

5. State what you want or need from the other person. "I want to know that you will keep your commitment to make special time for just you and me alone for at least part of the weekend we are together. It doesn't have to be doing anything special. I just need time with you alone." Often it is enough to tell the other person what you want without adding an ultimatum. Sometimes just having the problem brought to his attention is enough. But if you do give an ultimatum, make sure

you say what you mean, mean what you say, and do what you said you were going to do. Don't make an idle threat or one you are not willing to carry out. "If your friend is going to be with us all weekend, I'd like to take a break and spend the weekend with Mom and Grandpa" is more likely to be well received than "If she is going to be with us all the time from now on, I'll get Mom to go to court so I don't have to spend any time with you. You don't care anyway."

6. Be open to the other person's perspective on the situation. Give your parent the opportunity to talk, and be willing to listen. Perhaps you two had different expectations, or he never gave it any thought, or maybe he felt you would like to meet his friend. Often in assertive confrontation the problem is solved at this stage. Dad agrees to do what has been asked because it makes sense and is fair.

7. Negotiate an agreement you both can accept. If there is a difference of opinion or a disagreement about your proposed solution, you will need to negotiate a solution you both can accept. Decide on a time when you will get together to discuss how the proposed plan is working.

If assertive confrontation feels awkward to you, it will be difficult to teach to your children. During a divorce, there will be lots of opportunities to use this tool. Practice the steps in front of a mirror or with a trusted friend. Let your kids practice with you on issues that affect them. Since they won't have as many old habits to replace with the new, they might learn more quickly than you. You can begin to make assertive confrontation your tool of choice by recognizing your own feelings of hurt, sadness, frustration, or anger and by taking time to think before you respond, rather than reacting first and regretting later.

Even when you are using the seven steps, a confrontation can be frightening. Both you and your children need to recognize that to *keep* it a fair fight, it is always acceptable to:

- **Call time out.** "We're both too angry to talk right now. Let's talk about this later." "I'm too upset to work this through right now, I need to take a break." If either person is too angry or too upset to speak calmly and responsibly, it is important to call time out and come back to the confrontation at a later time.

- **Refuse to take abuse.** "It hurts when you call me a jerk." "Please don't attack my mom that way." "You can be angry at me, but you can't throw those barbs at me." If one person becomes verbally, physically, or emotionally abusive, the other person has the right to refuse to take abuse and to get out of the situation.

- **Insist on fair treatment.** "I'm willing to be flexible on the schedule. I just ask that you do likewise." "I realize your family reunion comes when I'm scheduled to have the kids at my home. I can agree to a change. I would like to be able to take Sarah on a trip just before school starts." Fair treatment is not always equal or identical treatment, but it is *honest, adequate,* and *just.*

We need to help our children learn to recognize when it is necessary to call time out. They need to know they have the right *not* to be verbally or physically abused by anyone and the right to be treated with respect, dignity, and fairness. They also have the responsibility not to let their feelings take over a situation; to see that they do not physically, verbally, or emotionally abuse another person; and to treat other people with the same respect and dignity they themselves ask for. Assertive confrontation enables parents and kids to use their feelings as a positive energy source to establish and maintain productive relationships with one another and with others outside the family.

Throughout the divorce proceedings, we can keep our cool in tense situations without putting our feelings on ice. Our feelings are real and legitimate, and we have the freedom to choose what we do with them. If we honor our feelings, our children's feelings, and their other parent's feelings, assertive confrontation can prove to be one of our greatest allies in the resolution of disputes that inevitably will arise.

Conflict resolution is essentially a process of bringing a submerged issue out of the darkened waters and up to the surface where it can be seen. Once exposed to the light, conflict can be a powerful teacher. It tells us what we don't want to hear but need to know.

—Brian Muldoon, *The Heart of Conflict*

Peaceful Resolution to a Violent Marriage

Relationships that do not end peacefully, do not end at all.
—Merrit Malloy, *The Quotable Quote Book*

Note: Throughout this section, men are identified as the violent partners and women as the abused partners. This is because most reported violent relationships involve a violent man who abuses his partner and their children (or her children). This is not to deny that there are women who abuse men, women who abuse their children, and men and women who together abuse their children. It is the effect on children that I am concerned about here, regardless of the gender of the person who commits the violent acts.

✻   ✻   ✻

Violence in a marriage is related to issues of power and control that have nothing to do with healthy relationships and interactions. If you are experiencing physical violence, emotional abuse, intimidation, coercion, or isolation, it is time to get help. In his book *The Gift of Fear*, Gavin de Becker makes this provocative statement: "The first time you are hit, you are a victim; the second time you are a volunteer." Sounds harsh, yet in that statement is a profound truth. If you can *choose* to take the second hit, then you are free *not* to take that hit; you do have a choice to get out of the situation. It is the choice in the matter that can set you and your children free. Believing that you deserved the hit, caused it, or can't get away from it leaves you and your children in a horrendous cycle of violence, fear, and guilt.

Children need to see that violence will not be tolerated and that something can be done to stop it. They need to know that there are constructive solutions to serious problems and that there are adults willing to help create these solutions. Battered parents and children cannot do it alone. If our concern is for the children, then we as a society must do everything possible to minimize the physical, emotional, and financial risks a family is exposed to when one parent chooses to get out of a violent relationship.

There are two rationalizations that can interfere with a peaceful break from a violent marriage:

1. **A false feeling of normal:** The abuse patterns have become such an ingrained part of the relationship that the violence is viewed as just another everyday occurrence. After you leave the relationship, it could be tempting for you to return to the security of the familiar, even if that familiar is violent.

2. **Justification and rationalization of the episodes of violence:** "If I'd kept the kids from crying, he would not have hit me." "He's really sorry for what he

did. Look at these beautiful flowers he gave me." "He just lost it. He didn't mean to do it." You will need help making and reaffirming your commitment to yourself and your children: "I don't deserve this, never did, and never will."

As well as the two rationalizations, there are three major concerns that an abused spouse must deal with when making the break: fear, finances, and family.

Fear of Retaliation

Believing that he owns his spouse and has a right to restrict her activities, the abusive partner has a tenacious hold on her, so the very act of leaving can be the most dangerous action the abused spouse can take. Without a support system in place to protect you and your children, no amount of legal assistance will help you. Protective orders alone help you if you have a reasonable spouse, but you wouldn't be fleeing if your spouse were reasonable. You need both a support system and legal protection to feel safe and begin to start a new life with your children.

What to Do When Love Turns Violent, by Marian Betancourt, gives detailed advice and good resources to help you and your children throughout the painful and frightening escape from a violent marriage. If you are ready to get out of a violent marriage, it is critical that you not do it alone. You will need help to develop a plan to leave, a plan to avail yourself of the resources in your community, and, most important, a plan to protect yourself and your children now and after you have actually made the break. Possible resources to get such help include, but are certainly not limited to, an abuse hotline, the police, and your family doctor. They then will be able to help you get the legal and professional assistance you and your children will need.

Financial Hardships

Financially unable to support herself and her children, a woman might see no way out of a violent marriage. Often the abusive spouse has full control of the money and credit cards. Unless there is a safe haven to escape to, an abused spouse is likely to choose a violent home over homelessness and abject poverty for herself and her children. The same support system that can help you leave can also provide counseling for you and your children and provide assistance in obtaining long-term housing and job training.

Family Breakup

The family must indeed "break apart" in order to heal the individuals in it. An abused spouse will likely need to create a strong family unit apart from the violent spouse. There is no coparenting plan on the horizon.

In response to an article written in *The Globe and Mail* in support of the two-parent family, in which editor in chief William Thorsell wrote that "other things being equal, single parenthood is not good for children," Anne Fisher wrote back, "Other things being equal, single parenthood is not good for parents either. But the reason many couples separate is that other things are not equal. . . . [O]ne sees women leaving their homes with children in tow, going to stay with family or friends or to crowded shelters. These parents face real hardships expressly for the sake of their children, to get them away from abusers, alcoholics, drug addicts or just a home that has become a battlefield."

If you and your children have been exposed to years of violence, leaving now will be difficult, and the repair work all of you will need to do will take time and a lot of effort. If it is the first hit, and your children have not witnessed the abuse, it is tempting to minimize the hit, make excuses, blame yourself, and be grateful that it is only you being hit, not the children. But one hit is too many, and the violence is not likely to stop. In fact, it is more likely to escalate. Leaving seems like such a drastic move, and it is. It is a difficult journey, one fraught with seemingly endless roadblocks and hazards. But the hazards to children living in a violent home are far greater.

The way married partners behave teaches ethical standards to the next generation. . . . Without the moral stability of a code of honorable behavior, children grow into adults who cannot create stable lives for themselves.

—Caroline Myss, *Anatomy of the Spirit*

Ages and Stages in the Violent Home

Much is made of the accelerating brutality of young people's crimes, but rarely does our concern for dangerous children translate into concern for children in danger. We fail to make the connection between the use of force on children themselves, and the violent antisocial behavior, or the connection between watching father batter mother and the child deducing the link between violence and masculinity.

—Letty Cottin Pogrebin, *Family and Politics*

Children in violent homes often suffer anxiety, fear, helplessness, grief, and physical injury in the short term and behavioral and emotional problems in the long term. Both the age of the children and the duration of the abuse influence how they will respond to the violence between adults in the home and what long-term effects that environment will have on them. By and large, all children from violent homes lose a part of their childhood and often find it difficult to fit in with their peers. Children who witness violence or are violated themselves in their homes are more anxious, frightened, worried, depressed, angry, distracted, and sullen than children from homes where there is no violence.

Infants

The greatest risks to the infant are neglect and Shaken Baby Syndrome. When the mother is stressed, anxious, or injured, her ability to care for her infant is seriously compromised.

Infants might respond to the loud noises and emotionally charged environment with irritability, crying, fitful sleep, and marked changes in eating habits. When the mother is distraught and the baby is irritable, the risk of physical harm to the baby increases. If the mother must keep the baby quiet for fear of further agitating the father, the baby's needs are ignored or smothered.

Toddlers

Actively involved in exploring the world around them, getting into things they are not supposed to get into, toddlers are at risk of being "slapped around," punched, threatened, or isolated for long periods of time. They might be denied food and drink or made to commit humiliating acts of contrition. They are at greatest risk of harm during the toilet-training process if accidents are dealt with punitively. Practicing their newfound language skills, they repeat what they hear: "Shut up," "Sit still or I'll hit you," "I'll kill you." They will act out their fears and anger by behaving aggressively with peers or by whining and clinging to any adult who offers them any kindness. They might appear anxious or sad.

Preschoolers

Trying to figure out who they are as unique, independent individuals, preschoolers look to adults for guidance, affirmation, and modeling. In a violent home, all three are distorted. Guidance sounds more like commands; affirmations are related to being quiet, not being seen, getting out of the way. Hugs are demanded or refused: "Give your dad a hug now!"

"I don't want a bad boy like you to sit next to me." Violent language and violent behavior witnessed in the adults, as well as the corresponding stifling of emotions and submissive behavior, are acted out over and over again in play. Children can't understand what is happening, but they can't let it go until they can make some sense of it. Dolls get smashed, stuffed animals torn apart, blocks pounded as children try to master their fears.

Five- to Nine-Year-Olds

Old enough to figure out that their home is different from other kids' homes, kids at this age are ashamed or afraid to bring friends to play at their house. If the violence is public knowledge, neighbors might forbid their children to play with the children from the violent home. This tends to victimize and isolate them even more.

These kids might stay home to protect a younger sibling. They might be fearful, mistrustful, and depressed. They might act aggressively toward their classmates, getting in trouble for doing or saying what is done and said every night at home. They lash out in an attempt to control their own fears or control someone or something in their life. They fantasize or create imaginary worlds to deny the reality of the violence happening around them and to them. They create intricate stories to cover up abuse or make their family life seem better than it is: "I bruised my arm when I fell out of bed" or "My mom can't help at school because she is packing for our trip to Disneyland." As they attempt to deny the reality of the violence, they repress their real feelings of sadness, rage, helplessness, and fear.

Preteens

Just as they should be developing a strong sense of self, preteens find themselves overwhelmed by the trauma of the violence they witness or experience firsthand. The feelings of anger and rage that went underground during childhood explode. Violated preteens are likely to be delinquent or victimized by delinquents, seeking out a target or setting themselves up as one. The imaginary world that allowed them to deny the pain and hurt at home now allows preteens to view the rest of the world in a distorted fashion. Without the love, affection, and respect that is their due as children, their hearts are empty, and empty hearts can be filled with anything, including bigotry, prejudice, and hate.

Adolescents

Able to reason as an adult and think things out rationally, adolescents are no longer able to buffer their pain with denial and fantasy. Since they can no longer deny the reality, they might try to minimize its effect on them, simply deny any pain associated with the violence, make excuses, or use humor to make the stark reality of the abuse seem less than it actually is. "Yeah, my dad strapped me, but I was asking for it." "Mom didn't mean to break my arm. I shouldn't have tried to get away from her." "It doesn't hurt, and anyway, now I've got a great excuse for not taking the exam." "Dad held a gun to her head, but he would never really hurt her." "She was drunk. She's really a great person when she's sober."

When feelings are denied or minimized, the heart that was empty in the preteen years now becomes a solid block of ice. If they don't let themselves feel the hurt, pain, and rage, adolescents are less likely to see themselves as worthy of compassion or to empathize with others who are suffering.

Teens who grow up in a violent home are at high risk for depression and self-destructive behaviors such as suicide, drug abuse, and self-mutilation. If the rage is not turned in on themselves, it is expressed outwardly in antisocial behaviors. Accustomed to living in crisis, some of these young people see high stress as a way of being; they need the adrenaline rush to feel

alive. They take dangerous risks without considering the consequences. Others adopt a "poor me" attitude: "I grew up in a
violent home; that's why I do the things I do. It's not my fault.
I have a good excuse for why I am the way I am." Others
become excessively needy for affection and possessions. Some
adopt the attitude of entitlement: "The whole world owes me,
and I'm not going to let anyone forget it."

*Generally, it seems the children whose families are in disarray or are in some
way dysfunctional are most at risk from factors outside the family. For example, children who face domestic violence are the most vulnerable to the
effects of growing up amid community violence. They are the ones who show
the most psychological problems and model the aggression they see else-
where. The experience of violence at home does not make these children
immune to the violence outside, it makes them more vulnerable to it.*

—James Garbarino,
Raising Children in a Socially Toxic Environment

In the Long Run
*Each person grows not only by her own talents and development of her inner
beliefs, but also by what she receives from the persons around her.*

—Iris Haberli

The impact of the violence doesn't go away when older teens
leave home. They carry the years of violence as excess baggage
with them. Unless the pain such violence produced is acknowledged and dealt with, this next generation of adults risks passing
the same pain on to their spouses, their children, and society as a
whole. The anger, fear, humiliation, and helplessness they experienced while being assaulted by a violent parent or watching
one parent assault the other recycles itself as they act out the
violence and tragedy all over again.

Clearly this is not the case for every child who grows up in a
violent home. In the Violence and the Family project of the

American Psychological Association, psychologist Jacquelyn Gentry found that "some children have a resilience" that buffers them from the effects of the violence around them. One of the most effective factors contributing to this resilience is "a person outside of the family who can be relied upon," be it a teacher, youth minister, family friend, or grandparent. Other factors can contribute to this resilience:

- Psychological "hardiness"—an innate ability to resist negative factors in the home
- Exposure to more positives than negatives in the family
- Development of strong self-esteem and strong social skills
- Good peer relationships
- A sense of hope
- High maternal empathy and support
- Opportunities to help others
- Respect for others, empathy
- Hobbies and other creative pursuits to find refuge in
- The development of some sense of control of one's life

To this list I would add a pet. A pet can prove to be an abused child's best friend. It might be the one source of unconditional love that the child has. And caring for the pet can give a child an opportunity to get outside herself, outside the pain, and be playful.

No one should condone any kind of violence against children, and every effort needs to be made to prevent the violence in the first place. We simply cannot assume, however, that all is lost, that a child is doomed if she has been a victim of abuse. Children can and do fight the odds and grow up to be ardent advocates of children's rights, themselves refusing to repeat the abuse they knew firsthand.

My satisfaction comes from my commitment to advancing a better world.

—**Faye Wattleton**

Moving On

Family is content not form.

—**Gloria Steinem**

Once the steps have been taken to move out of a violent marriage, the healing can begin. The TAO of Divorce is as relevant in the peaceful resolution of a violent marriage as it is in the dissolution of any other marriage. Kids will need your time, your affection, your sense of optimism, and the six critical life messages. And you will need to practice restraint, kindness, and intelligence.

Most likely your children have witnessed the violence you are all leaving behind. Yet they still need to feel some kind of connection to the other parent. This connection might be through pictures, letters, or "to be saved and sent later" life-story books.

In the beginning, and perhaps for a long time, it might be necessary to limit the contact between your children and their other parent. Seek the help of a therapist who specializes in trauma and structured/supervised visitations to help you do what is ultimately best for your children.

Striving to maintain a connection with the absent parent isn't just a matter of making sure your children don't lose touch; it's about helping your children see their parent as a whole person. They need to hear about the good traits (think hard, you saw good traits in your former spouse once), the good times (this might be a bigger stretch for you, but so important for your children), and the goodness underneath the violence. No person is all bad. Kids need to know that this person who is their parent is a person who has made poor choices. They don't need to repeat those choices; they can take the good traits they might have inherited from their parent and choose to make good choices with these traits.

When it comes to the violence, what to tell your children will create a dilemma for you as you struggle with telling the truth with restraint, kindness, and intelligence. The Sufis have a saying that can serve as a guide: "Your words must pass through three gates: Is it true? Is it necessary? Is it kind?" Something might be true but not necessary to say, or it might be true and necessary but come out dripping with vengeance or sarcasm. If you aren't saying what you need to say in a kind way, go back through the gates and start over. "Your father is a no-good, lying, cheating, violent, drunk bum" won't even get through the first gate. "I know you would like to see your dad on the week-end, but his hitting people when he is angry makes it unsafe right now for you to spend time alone with him." You can also tell your child, "He is getting help to solve his problem so that he can spend time with you in the future," if such a statement is true. If it's not true, don't say it. (If the parent has abandoned the children or has terminated any kind of contact, children need to be helped to understand that it has nothing to do with them and everything to do with the fact that either the parent needs this time right now to take care of himself and get his life back on track or that there is something seriously wrong with someone who cannot love his children.)

It might take a long time before any semblance of a normal relationship between your children and their other parent can begin. The violence will not stop until the abuser is willing to take full responsibility for what he has done, speak of his pain that has been driven underground, and make a concerted effort to learn new conflict-resolution skills. This means that the abuser must face the ways in which he was hurt, face his guilt and shame for hurting others, and acknowledge that he was solely responsible for his own violent behaviors—no excuses.

If the abuser is willing to take these necessary steps, it is possible to have a peaceful resolution to a violent marriage that includes a recommitment to the original vows (rare) or a new commitment to a coparenting plan in two separate homes (difficult but not impossible). In order for this to happen, the abused spouse and children need to be out of the violent relationship and out of harm's way as the abuser takes the necessary therapeutic steps to "fix himself first" before attempting what in reality would be a *new* relationship with both his spouse (or former spouse) and his children. The abused spouse will need to work through her own issues related to the abuse and what she might or might not have done to contribute to the violent environment in the home.

All this said, sometimes the only peaceful resolution that can be had is for the abused parent and her children to create a strong home together, apart from any involvement with a parent who refuses to get help and continues to behave violently. When this is the necessary decision, the abused parent can do all of the above for the children and go one step further for herself: refuse to condemn the abuser. To condemn him is to get caught in the negative energy of the condemnation. This negative energy acts as a black hole in your universe, swallowing your own energy and optimism. Once caught in the hole, it becomes nearly impossible to pull yourself out of it and get on with your life. Deny the condemnation its power.

When you are offended at anyone's fault, turn to yourself and study your own failings. By attending them, you will forget your anger and learn to live wisely.
—Marcus Aurelius

Chapter 5

Families Born of Loss and Hope

[Families have] weathered combinations of step, foster, single, adoptive, surrogate, frozen embryo, and sperm bank. They've multiplied, divided, extended and banded into communes. They've been assaulted by technology, battered by sexual revolutions, and confused by role reversals. But they are still here . . . playing to a full house.

—Erma Bombeck, *Family: The Ties that Bind . . . And Gag!*

For years, the definition of "family" has been debated and batted around by many diverse groups. The debate is often tainted with moral absolutisms and ideological biases, and the definition generally ends up being more exclusive than inclusive. To get hung up on what is the best kinship structure is to deny the variations that have served us well throughout history, in different times and places, in different cultures, and through great chaos and loss. To define the family as only one type of kinship structure is to define others as aberrations or deviations, not "real" families.

Stepfamilies, foster families, single-parent families, and adoptive families are no more and no less "real" than the one that is traditionally spoken of as the "family of origin," "natural family," or "biological family." The three characteristics of family hold true for a family of one child and one parent; two biological parents and two children; two stepparents with his, hers, and ours; parents and adopted children; grandparents raising their children's children; four generations living together; and any number of variations on each of these combinations. Professor Richard McKenzie, author of *The Home: A Memoir of Growing Up in an Orphanage*, would argue convincingly that the orphanage he grew up in was his real home and the people he lived with his real family.

A family is:

1. A small group of people bound together by commitment, caring, cooperation, traditions, common rituals, and common language, whose boundaries are identifiable, firm, and flexible.
2. A microsystem in which each member is affected by and affects other members and the system as a whole.
3. Part of a larger community (macrosystem) that interacts with and influences the family and its individual members (social, economic, and political).

A new baby is born, a teen goes off to college, a father dies, a grandparent moves in, a daughter marries, a foster child is adopted. All families form, grow, shrink; they change in structure, function, and interaction through the marital, parental, and kin relationships that are integral to the family system. These relationships are by their nature dynamic, interconnected, interactive, and reciprocal. The family in its essence continues and constantly redefines and renews itself.

The three types of families discussed in this chapter are all born of some kind of loss and hope. Children born into a single-parent family or who are a part of a single-parent family as the result of a death or divorce, children entering into a stepfamily or being born into one, children who are adopted or who are created through new reproductive technology—all will have as a part of their life story any number of physical, emotional, and psychological losses that are associated with their relationship to a family of origin. The new family they are now a part of is created not only with the threads of loss but also with the vibrantly colored threads of hope and new beginnings. These threads of loss and hope are inextricably woven into the entire fabric of the new family. To deny the loss or to deny the hope is to deny the reality of these families. To view only the hope and new beginnings and negate the issues of loss is to deny children the opportunity to grieve their real losses. To emphasize only the losses is to negate the strength of the human spirit and the will to move beyond grief and begin anew.

As well, to view these families as problem families in need of a solution is to burden them with labels, stereotypes, false assumptions, negative bias, and prejudice that neither helps them function as an effective family unit nor allows them to partake fully in the larger community. Comments such as "He comes from a single-parent home," or "She's their adopted daughter" or "He lives only with his stepfather" are usually not comments merely describing a relationship; what is implied is the prediction that the children in those families are doomed to a life full of misery and possible jail time. Any hope is left out of the equation.

There are problems inherent in these families, but they are problems that can be solved when addressed without bias. They fall into five categories: loss, boundary ambiguity, communication, commitment, and discipline.

1. Loss
Death, divorce, infertility, adoption—the losses involved in each plunge adults and children alike into grief. This grief cannot be denied, refused, overlooked, minimized, or belittled. It must be named and faced head-on. If the family born of the loss is to thrive, the three passages of grief (piercing grief, intense sorrow, and sadness that shares space with joy and peace) need to be honored, not rushed or cast aside.

In families born of loss and hope, not everyone will experience the same three passages to the same extent as anyone else in that family, nor will everyone do it in the same order. Just as we find when facing a death, these passages are fluid and overlapping; they can all be present in the same day and can all barge into our lives unbidden and without notice.

The piercing grief of good-bye can be felt once at the time of a divorce and once again by a child at his parent's second wedding. Mom might be full of hope for a new beginning and child might be full of piercing grief, knowing there is no more hope of his parents' getting back together.

A never-before-married stepdad might grieve the loss of the "idealized" marriage to a never-before-married woman who brings no children with her to the altar. The grief can go hand in hand with the joy of marrying his true love, who has three children walking up the aisle with her.

Parents might be feeling the piercing grief of good-bye to any hopes of having a biological child as they also rejoice in the adoption of a new baby. The new baby doesn't know this grief and will enter his own passages of grief much later, as he comes to understand his adoption.

A child adopted after living in several foster homes and a violent family of origin will often feel all three passages on the same day, as everyone else is rejoicing over the good news of the final adoption papers. Now she will never go home again to her family of origin; she might feel relief, piercing grief, and intense sorrow, as well as the joy and peace of truly belonging in her new family.

A teen might be so happy to get out of the job of caring for her younger siblings in her single-parent family that her father's marriage leaves her excited and joyful. Years later, at her own marriage, she once again feels the piercing grief of her mother's death.

The third passage is a constant companion to each family member individually and to the family as a whole. When given its due, sadness tempered with joy and peace allows parents and children alike to speak openly of a death, a divorce, infertility, the longing of a birth mother without these consuming the present and defining the future. The keys are to name the loss, honor the grief, and confront the pain.

2. Boundary ambiguity

"Who's in and who's not?" "Who's the parent and who's the kid?" "Whose parent and which kid?" "What's appropriate, what's not?" These simple questions address the issue of family boundaries. In families born of loss and hope, the answers can be complex and conflicting.

In a stepfamily, it is critical that each family member has a sense of having a special place in the new family; yet often everyone in the family has an equally important place in a different family system, with another set of relationships: stepparent, stepsiblings, half siblings, grandparents, aunts, uncles, cousins.

When the family boundaries become so fluid as not to be boundaries at all, with people coming and going at such a pace that they never form a family unit, kids and parents suffer from lack of identity. When boundaries are so rigid as not to allow kids access to their biological parent or acceptance into the new family unit, kids suffer from lack of connectedness. A middle ground is difficult to achieve but necessary for the well-being of the family.

In a single-parent family, the internal boundary ambiguity can be more of an issue than the external. Boundaries are difficult to define if the child assumes the role or responsibilities of another adult rather than being able to be a kid; or the reverse, if the parent acts like a kid and the oldest sibling parents both the younger siblings and the parent. "You be the kid, I'll be the parent" is a helpful mantra for defining boundaries.

It is easy for divorced parents to fall into the trap of seeing a single-parent home as nothing more than a pause in between the first family and the stepfamily, rather than a functioning family unit in and of itself. The household is a hodgepodge of furniture, simply thrown together with no thought or desire to create a "homey environment" for the children. Pictures remain unpacked in boxes stored in various rooms to wait for the time when there is a "real" home to hang them up in. No thought is given to a special place for the children to put their clothes, toys, and memories. Meals are unplanned and on the run.

Creating rituals, routines, special places, common spaces, and common memories helps to create the external boundaries in both single-parent families and stepfamilies. These boundaries become tools to help children create some order in their lives, and with this order, move out into the larger world having a strong sense of a place to call home and people to call family.

In the adoptive family, the child has a connection to both her biological parents and her adoptive parents. Even in closed adoption, the connection to the biological parent is present in the psyche and genes of the child. Open adoptions have blurred the boundaries even more, with children often having close contact with a birth parent who is more like a visiting relative or friend than a parent.

On the horizon is the issue of boundary ambiguity in families created with the help of sperm donors, egg donors, surrogate parents, and other forms of NRT (new reproductive technology). In this arena there are no easy answers, but necessary questions that are just beginning to be asked. For example: If the child is created as the result of a sperm donation and the egg from the wife, is the husband the father? Or is it more accurate to state that the husband is in reality the stepfather or adoptive father, as the child has a biological "father" via the donated sperm? Does the child have a right to know his true biological heritage? Does he have the right to know that he has twenty or thirty half siblings somewhere? What if by chance he falls in love with a half sibling? (These are not new questions, just old

questions framed in a new science. Adopted children have won-
dered about siblings and half siblings. Cousins have discovered
that they were really brothers. Sisters have found hidden birth
certificates, only to discover that they are not even related.
Brothers have found out late in life that their older sister was
really their mother.)

If the child is created as a result of a donated egg and the husband's sperm, is the wife, who was pregnant for those nine
months, the mother or the stepmother or adoptive mother, as
she is not genetically related to the child? What biological relationship does a woman have to a child if she "rents her womb
for nine months" to a husband and wife, giving birth to a child
to whom the husband and wife are both genetically related?
What are the child's rights? Do these rights supersede parents'
wants and needs for privacy in such an intimate procreative
issue? How damaging are the secrets?

3. Communication

In order to relate to one another in a marital, parental, or kin
relationship, we need a relevant language and lines of communication that are available, open, and direct. Language helps define
who we are as individuals and in relation to one another. It is
difficult to tell our story when there is a paucity of language or
when the language is steeped in negative definition. We stumble
over the half sister, stepbrother labels and then try to explain the
biologically related sibling: Saying she is my "real" sister implies
that the others are in a less than "real" relationship. Even the
phrase "traditional family" implies that all others are nontraditional and second-best, even though, as Suanne Kelman points
out in *All in the Family,* historically the family has been more
often "nontraditional" than "traditional."

Lines of communication are too often severed in divorce,
convoluted in stepfamilies, sometimes open and sometimes shut

in adoption, shut down completely or shrouded in secrecy in the case of NRT. For any family to communicate well, it must be able to create its own history and its own stories around that history.

In *Cinderella Revisited: Surviving in a Stepfamily Without a Fairy Godmother,* Dr. Peter Marshall talks about the "Good Family Myth," in which stepparents feel the need for approval from society more than they feel the need to be honest, frequently defending themselves and their family to others, afraid to admit to problems, afraid to seek help in times of crisis for fear of being seen as "a poor imitation of the real thing." The opposite, the "Broken Family Myth," invites unwarranted criticism from society and unwarranted and unnecessary apologies from stepparents. With both myths, real communication takes a backseat to maintaining the façade or overcoming bias. The same is often true of foster or adoptive families.

4. Commitment

Creating a family takes time and commitment from all involved—commitment to one another and commitment to the family as a functioning entity in and of itself. Commitment involves certain rights, responsibilities, and obligations that go beyond, yet may be bound by, law (in the case of stepparents, adoption, and NRT, legal protection doesn't always exist).

An adult who helps create a stepfamily will have less interest in working through the inevitable rough times if she enters the relationship halfheartedly, with the belief that if this marriage doesn't work out, she can bail out, than if she makes a strong commitment to her marriage and her family at the outset. Some teens are committed to making their parents' second marriage fail and expend a lot of energy working toward that goal.

A single parent might view the single-parent family as only a phase to be gotten through before the next phase "rescues" him from the troubles of raising children alone, instead of committing to making it a functioning home with all the healthy dynamics of family—boundaries, routines, rituals, and traditions.

A child who has been adopted might worry that he will be abandoned by his parents if the parents run into rough times or if he does not live up to their expectations.

At present, commitment on the part of sperm donors and egg donors is not an issue, but parental responsibility in case of divorce in such situations is beginning to be challenged. If the father of a child is not the biological father because of a sperm donation, does he in fact have any rights, responsibilities, and obligations to the child he has raised? Can these be contested or overruled? And where does commitment fit into the picture?

5. Discipline

For the most part, in one kind of family, the discipline issues around mealtime, bedtime, sibling rivalry, chores, and getting your teen out of jail are the same as in any other kind of family. The need to have limits and boundaries, rules and consequences, and to follow through is no different in a family where all members are biologically related than in a family where some children are biologically related to the parents, some adopted, and some in foster care. The ages and stages are the same; the purpose of discipline (i.e., to give life to learning) is the same; the steps are the same; what is different is *history*—where each person involved in the process is coming from and what the history of the relationship between the parent and the child is at the present.

A child coming into a family from an abusive setting where punishment was the norm will have different expectations and/or fears when he drops a glass, or wets his pants, or argues with his sister than will a child who has come into the family from a setting where there were no limits and boundaries, rules, or consequences. And both will respond differently than a child who comes into the family from a setting where healthy discipline was the norm. As well, a parent accustomed to taking an authoritarian stance will run into tremendous resistance from both his new spouse and her children when he tries to "put his foot down" at the dinner table. A mutiny is likely to ensue, with the mate being thrown overboard. Teens don't care to have even their own biological parent disciplining them, let alone a new adult in the house issuing new rules and expectations.

Fearful of being rejected by his children who live with him during the summer, a father might let his kids run wild while the stepchildren he lives with all year are still expected to follow the rules and routines of the house. A stepmom might be expected to *do* for her stepchildren but never *say* anything when the kids are definitely out of line. Add to all of this the ghosts in the house: "But Dad lets us do that at his house" or "If Mom was alive, she'd let me" or "You're not my real parent." Then throw in all the different parents' discipline tools from their own families of origin. Of the five problematic areas, it is no wonder that this one is often the most obvious and the most divisive for families born of loss and hope.

These five areas of concern—loss, boundary ambiguities, communication, commitment, and discipline—are not unique to families born of loss and hope. All five can be issues in any first marriage with both parents biologically related to their children. My husband Don's first fiancée was killed in an accident. Her *loss* will forever be a part of Don's life story, and by extension a part of mine, and an even lesser part, but still a part, of our children's life story.

Being of Irish heritage, with few close relatives, I married into Don's extended Italian family. At weddings, christenings, and funerals, I met more Italian relatives than there were people in the small village I grew up in; talk about *boundary ambiguities*. To this day, I'm still not sure who's a cousin to whom and how, and how much they all know about one another—and who really cares?

We still deal with one another's spoken and unspoken *communication* within our own families of origin. "What? Not have spaghetti on Sunday?" "Of course we all talk at once at the breakfast table. Nobody would ever dream of reading the morning paper, and certainly not the obituaries." I send my parents a rose on my birthday. Gathering at Don's parents' for his birthday is a tradition not to be messed with. The differences between an Italian funeral in the cathedral and an Irish wake in a home was an eye-opener for both of us. My family opened gifts on Christmas morning, Don's on Christmas eve. The medals Gram insisted on pinning on our babies' shirts to protect them from the *malocchio* (evil eye) raised a few eyebrows on my side of the family. My being a former nun raised a few eyebrows on Don's side.

Commitment to one another was and is still a work in progress, but balancing that commitment with sometimes competing extended family commitments was a bigger struggle than either of us could have ever dreamed possible.

Don, a rehabilitation counselor, and I, a special-education teacher, had several years of practice when it came to *discipline* before we had our own children. A few of our practices were conflicting. And we carried in us the well-worn tools from generations past—some good and some not so good. Our three children, who are like day, night, and afternoon, compound all the issues even further.

No family is without its complications and complexities, its losses, boundary ambiguities, communication barriers, commitment concerns, and discipline hassles. Being aware of the ones unique to your kinship structure is step one in helping you avoid the pitfalls that can undermine your family.

Underground issues from one relationship or context invariably fuel our fires in another.

—Harriet Lerner, *The Dance of Anger*

The TAO of Hope

A new philosophy, a way of life, is not given for nothing. It has to be paid dearly for and only acquired with much patience and great effort.

—Fyodor Dostoyevsky

Just as any family that is born of loss and hope is, by its nature, complex and complicated, so is its TAO. To find the TAO of Hope, take the TAO of Family (time, affection, and a sense of optimism), add the three passages of grief (piercing grief, intense sorrow, sadness tempered with joy and peace), the TAO of Mourning (including the six critical life messages), and the TAO of Divorce (acting with kindness, restraint, and intelligence); multiply them all by the number of family systems involved, and add **compromise, generativity,** and **balance**.

Compromise, the act of adjusting or settling by mutual concession, will become second nature to you as time passes, if all adults involved in the various family systems are willing to embrace **generativity**. Psychoanalyst Erik Erikson coined the term to speak of the responsibility the older generation has to the younger. He defined it as "devoting our time and energy to the nurturing of the next generation as opposed to being fully absorbed in our own lives and pleasures." I use his term to express a critical element that all adults involved in a family born of loss and hope must factor into their decisions and interactions. To make such a family a safe harbor for children, the rights, needs, and wants of all parties need to be **balanced** against the final weight of what is good, just, and right for the children.

Wanting to create a new family devoid of any connections to

a former spouse might make your life less complicated, but it denies your child his right to develop a bond with one of his biological parents. Your want is superseded by your child's basic right. Compromise with your former spouse becomes critical to the nurturing of your child. Similarly, keeping an adoption secret so as to create the illusion of a biological family might make it easier on the adults involved, but it denies a child her right to know the truth about her own biological and genetic heritage. It could create much more havoc in the lives of everyone involved should the secret necessarily or unwittingly be disclosed after you've been living the family lie for years.

As you and your family navigate the turbulent waters of the five problem areas, know that you can all get through each one of them by underpinning your resolve with the TAO of Hope.

time, affection, sense of optimism

+ piercing grief, intense sorrow, sadness

tempered with joy and peace

+ six critical life messages

+ kindness, restraint, and intelligence

+ compromise, generativity, and balance

= the TAO of Hope

For me, it is far better to grasp the Universe as it really is than to persist in delusion, however satisfying and reassuring.

—Carl Sagan, *The Demon-Haunted World*

Adoptive Families

Oh what a tangled web do parents weave / When they think that their children are naive.

—Ogden Nash, *Baby, What Makes the Sky Blue?*

In the early twentieth century, it was common for a family friend, neighbor, or relative to raise a child who might or might not have contact with her biological parents. Family boundaries were often quite fluid and functional. My cousin's grandmother, Nellie C., was raised by neighbors after her parents died during the flu epidemic in the early 1900s. Throughout her ninety-seven years, she held dear her brothers and sisters—once neighbors and playmates—who together grew to love one another as family. And they were family in every way: loving, argumentative, and affectionate.

Other informal arrangements could take place within the boundaries of one's extended family. Children were sent overseas to live with more affluent or educated relatives. Dying parents bequeathed their children to family friends. Unable to care adequately for all her children, a mother might "gift" one or more of them to her sister, with siblings then raised as cousins. Secrets were sometimes kept: An older sibling might in reality have been the birth mother of a youngster raised to believe that she was his sister. There was little if any involvement by governmental agencies.

In the mid-1900s, adoptions became more formalized, with parents and children connected through doctors, lawyers, and social workers. There was a concerted effort to match the physical characteristics of the parents and children so that they could be seen as a biologically connected family. These were often "secret" adoptions: Both the birth mother and the adoptive mother would "visit" relatives for a period of time; then the birth mother would return home with no further mention of the baby, and the adoptive mother would return home having "given birth" while away.

By the late 1940s, with an emphasis on the primary importance of the "nuclear family" and the stereotyping of such families through the media, the secrecy, stigma, shame, and institutionalization of adoption became entrenched through state and provincial laws. Birth records were sealed, and new birth certificates were issued listing the adoptive parents in place of the biological parents. This was to be known as a "closed" adoption. Contact between the birth parents and the adoptive parents was deemed unnecessary and unhealthy. The secrecy, intended to protect the privacy of the adults and children involved, kept these same people in the dark about one another.

By the 1960s, children adopted under these statutes were reaching their late teens and early twenties. Some began to search for their biological parents, often with the support of their adoptive parents. Biological parents, many of whom felt "forced" to place their child for adoption because of their age, family shame, cultural taboos, or societal pressures, began to seek information about their child's well-being or seek contact with the child.

By 1980, both sides of the adoption equation were changing. On the front end, "open adoptions" were being tested both formally and informally. Who could adopt and who couldn't was being challenged. Interracial adoptions were being proposed and opposed. News accounts of children in need of adoption in Poland, Romania, and China opened one more avenue for adoptive parents and allowed some to escape the challenges of "open adoption," only to replace them with the different challenges of "international adoption."

On the back end, records were being unsealed. Adopted children and their biological parents could choose to register with various groups, hoping to make contact with one another through a mediator. Accounts of reunions, some good and some not so good, flooded the media. Not only were the pros and cons of open records being debated, the debate was playing out in living color in many living rooms. Secrets, stigma, shame, silence, hidden photos and letters, liaisons concealed were like skeletons rattling around in a closet for far too long, finally pushing open the door and falling out onto the floor for the knowing and unsuspecting alike to see. Wives told husbands of unknown children, fathers told sons and daughters of siblings long hidden. Siblings met brothers and sisters they never knew existed. Adoptive parents helped or hindered the search, their own fears and anxieties taking center stage or given second billing to their adopted daughters' and sons' search for their biological roots. They were there for the joy, the elation, the tears, the letdowns, the dreams that did not have fairy-tale endings, and the realities of a now "open adoption." Issues faced by stepfamilies for generations were now spotlighted for all parties concerned: titles, names, histories, holidays, traditions, rituals, graduations, weddings, visits, grandchildren, and inheritance.

Today, with as many variations on the theme as existed in the early 1900s, plus the added elements of transracial adoptions, multicultural adoptions, international adoptions, single-parent adoptions, and varying degrees of open adoption, the adoption process is forever changed. But the emotional, physical, and psychological needs of the child who was adopted have not.

When we deny adoption's losses, we also deny ourselves its fullest blessings.
—Marcy Wineman Axness

The Language of Adoption
What you teach your children is what you really believe in.
—Cathy Warner Weatherford

Both you and your child will probably need to deal with negative comments from adults and children who hold strong prejudices about family and do not understand adoption. Callous questions and comments hurt. "How could your real mom just give you away?" "Do you know who his real parents are?" "Why couldn't her parents take care of her?" As much as you would like to respond in kind with a callous comment, don't.

Take the opportunity to show your child that such an insensitive gift does not have to be accepted. As long as it is not accepted by the receiver, it remains with the giver. A simple comment like "I am his parent" should suffice to get the message across. If the question is particularly insensitive, you might answer with an assertive response: "Why would you ask such a personal question?"

Just as language is key to the messages conveyed when speaking about divorce, remarriage, and stepfamilies, callous stereotypes and hurtful words used to speak about adoption can isolate parents and their children, brand them, undermine their sense of self, family, and parenthood, and connote second-class status worthy of pity or condescension or worse, condemnation.

Simply by changing hurtful terms, we can improve our own and our children's sense of identity, sense of family, and sense of belonging in an extended family and in community. *Adoptive Families* magazine has published a list of terms considered hurtful to participants in adoption, along with recommended alternatives:

Old Language
- real parent, natural parent

- adopted child, own child
- illegitimate
- give away, place for adoption
- reunion

- adoptive parent
- search, track down parents
- an unwanted child
- hard to place

New Language
- birth parent or biological parent
- my child
- born to unmarried parents
- make an adoption plan

- meeting, making contact with
- parent
- to locate, contact
- a child in need of adoption
- child who has special needs

Old Language	New Language
• foreign child	• child from another country
• is adopted	• was adopted

The language doesn't deny the reality of the adoption, nor does it eliminate the loss, the hurt, and the pain that all the parties to the adoption must deal with because the adoption happened.

"I *was* adopted" instead of "I *am* adopted" allows a child to acknowledge that the event happened and it is a part of her past, but she need not let it define her entire life or put a stranglehold on her future. She *is* Susan, and she *was* adopted. "Gave away" not only conveys a selfish birth parent but also presents the child as a disposable commodity. "Made an adoption plan" conveys that thought went into a difficult decision. What we call ourselves and what we are called can greatly influence who we become.

Each person grows not only by her own talents and development of her inner beliefs, but also by what she receives from the persons around her.

—Iris Haberli

Ages and Stages

Adoptive kids often grow up with the mantra "Being adopted is just another way to become a family." This is a dismissive characterization of a profound experience which has involved not only the parents' deep losses, but the child's loss of parents who couldn't keep him. With the best of intentions, adoptive parents often convey half-truths about the implications of adoption to shield their child from the pain of loss which is inherent in the experience.

—Marcy Wineman Axness

A child's adoption will influence her emotional, psychological, and social development. How you as a parent deal with the issues of loss, grief, confusion, anger, and identity can aid or hinder her development in all these areas. Sometimes you will need to guide and speak, at other times watch and listen. Open, honest communication, a willingness to go over and over the issues as your child gradually absorbs the complexities of adoption, and an acceptance of your own grief and your child's grief will help your child grow up with a strong sense of self, a strong sense of family, and the knowledge that she is loved.

Throughout the various ages and stages, your child will learn to attach and to separate. These two actions cannot be viewed as distinct from each other, since the interplay between them creates a whole. A dynamic balance between the two are necessary for a child to become a capable and independent person who can embrace both his aloneness and his being in community, his ability to act independently as well as relate intimately.

The tension between separation and attachment is often amplified in a child who is adopted. There can be a tendency to hang on to others for fear of being abandoned or to isolate himself so as not to risk hurt and rejection. This tendency can be minimized if you and your child can develop a strong bond, trust one another, and communicate openly and honestly.

Note: The ages and stages for adoption are similar to the ages and stages for a child in a single-parent family, two-household family, stepfamily, or any other family constellation born of loss and hope. The issues around separation, attachment, loss, boundary ambiguity, communication, commitment, discipline, and the existence of a birth parent (whether down the street, uncommunicative, unknown, or dead) will need to be addressed in all these kinship structures.

Infants

In this first year of life, your baby is trying to "make sense" of the world around him, bonding with primary caregivers and developing a trust in the world as a caring and responsive place to be. The concerns in infancy are attachment, security, and predictability.

TAO of Hope: Provide loving, consistent care, nurturing, safety, and opportunities to explore. Respond to infant's needs with gentle touching, talking, and singing. Keep changes in daily routine to a minimum. Infants can't comprehend any concept of adoption, but they will respond to changes in their environment, including a change in the primary caregiver.

Use this time to deal with your own feelings of grief and loss as you also celebrate the joy of having this baby to care for and love. Soon enough, you and your child will be thrown into the passages of grief together. Relish this time of calm innocence in your child's life as you face the realities of losses that your baby's presence amplify and at the same time lessen. The loss might be your inability to bear your own children, or it might be the loss you feel that this child will not share with you and possibly your biological children the familiar smile or curly hair, dimples, or shoe size. It does no good to feel guilty about feeling grief; honor the grief, name it, and make peace with it.

You can begin a storybook about your child's life with anecdotes from all family members, pictures of the family before she arrived, the day of the adoption, pictures of her birth family, and events in her life after the adoption.

Toddlers

Separation and attachment activities intensify at this age, so your toddler might show a strong reaction to any change in his living situation. Your toddler will normally show signs of anxiety when you leave the room and at the same time protest loudly when you want to hold him and he wants to get down. The verbal and physical expressions of anger and aggression, the defiant nos, attempts to do things by himself—these are typical behaviors for all toddlers.

Although his verbal communication might be limited, his ability to comprehend what is being spoken is more developed. He might overhear adults talking about his adoption and not only understand some of the words but understand the message conveyed by your tone of voice and emotions expressed.

TAO of Hope: Let your toddler know in lots of ways that he is loved, protected, and understood. Although your child won't ask many questions at this age, you can begin to tell him in stories about his adoption, show him the storybook you have been creating, and be prepared to tell the stories and show the pictures over and over again. He will not yet understand the term "adoption," but he will be able to begin to develop a positive view of adoption if his story is told in a warm, loving family environment. The key elements are simplicity and repetition without a lot of confusing details. Forget your painful search, detailed paperwork, your fears, and your frustrations—your child doesn't need to hear them now.

Be aware that your child might be playing and at the same time listening to the adult conversations going on around him. Check out what is being said and how it is being expressed. Is the message one you want your child to get?

Preschoolers

Busy establishing their individual identity, learning new motor and linguistic skills, figuring out roles and power relationships, making friends outside their family, and realizing that parents don't have answers to all their questions, preschoolers are ready to begin exploring how they became a part of the family. Your preschooler uses "magical thinking," believing that her thoughts, words, and actions to have great power.

She is trying to separate fantasy from reality. She asks questions about her body and the world around her. She might be very verbal, but her ability to verbalize her feelings will be lagging behind her body's experience and expression of feelings. The "wow" and "why" of life that renowned scientist Carl Sagan called the "gift of wonder and skepticism" are being expressed with fervor. "Wow, look at the sky!" "Wow, look at the spaghetti!" "Wow, look at this mud puddle!" "Why do I have to?" "Why not?" "Why is her hair red?" "Why is your hair curly and mine straight?"

Birthdays take on a special significance at this age, and not just because of the party and gifts. As your child comes to recognize her own uniqueness as an individual, this day will stand out as her special day.

Whether or not the birth parents are active in her life, the issue of boundary ambiguity comes into play. Who belongs where? Who is called what? Where do I fit? The presence of the birth parents can create conflict and blur lines, but the flip side—knowing who they are and why they made their choices about the adoption—can take away unnecessary fears, fantasies, and mystery.

TAO of Hope: Take advantage of the nonstop questions to begin to explain where babies come from, how they are born, and that all children are born but not all children are adopted. Mom didn't give birth to her; her birth mother did. Hair color and skin color come from birth parents. You might not be able to answer all of her questions completely, but if you try to answer as many questions as you can, forthrightly and honestly, she will be more likely to keep the questions coming. Don't ignore any questions she has about her adoption or try to change the subject. At the same time, don't let the topic of adoption be the overriding concern all the time, every day. It is a *part* of her life story, it is not her *whole story*.

Continue sharing her life book with her, and let her add

anecdotes, pictures, handprints. Storybooks about adoption can be incorporated into the library you share with her. Books such as *Adoption Is for Always* and *William Is My Brother* can help your preschooler understand her own adoption and begin to understand that she is not the only person to be adopted and that her place in the family is forever.

As your preschooler begins to understand the concept of adoption, she will begin to feel the pain and the sense of loss that come with it. It is important that she sees you expressing a wide range of emotions and that you encourage her to express her sadness and her grief as well as her excitement and happiness. Know that birthday parties are a time of both joy and sadness for you and your child. Give her the opportunity to "connect" herself back to her biological parents through pictures and storytelling on this day, and celebrate with her her own uniqueness as a one-of-a-kind person, separate from everyone else. You can also celebrate the special day she was adopted, your "Gotcha Day." All kids love a celebration, and what a great day to celebrate together.

Five- to Nine-Year-Olds

At this age, your youngster tends to listen in order to collect information, compare, test, and disagree with adults and peers. He might ask many questions about the specifics of his adoption, cast judgment on his birth parents, tell you he wished he wasn't adopted, and argue with you about your notion that adoption is a good way to help create a family.

Challenging your values, arguing with you, and hassling over minute details of family rules and routine, he seems to be pushing away from the family, but at the same time he is fearful of being abandoned. The near-hurricane force of the separation/attachment struggle is as exhausting for him as it is for you. A need to be a unique individual, while feeling that he truly belongs in the family, is a struggle carried out at the breakfast table in the morning, in his mind during the day, and in his dreams at night.

He also imagines things about his birth parents, wondering out loud, if he feels safe talking with you, about any and all aspects of his adoption. If he suspects you are upset, angry, or hurt by his musings and questions, he might stop asking and keep his fantasies and fears to himself. Facts and feelings might go underground and fester until anger and aggression erupt in the schoolyard. Once a delightful tale about coming into the family, the adoption story he has heard since his preschool years takes on an ominous tone in his head. If this family welcomed him, someone else had to *not* welcome him, for whatever reason. Was he bad? Was he not good enough? Whereas in the preschool years he felt a sadness, he is now intellectually able to begin grappling with the various implications of his adoption. Both piercing grief and intense sorrow are felt. When he was younger, he thought you could fix all his hurts; now he must come to grips with the fact that you are powerless to protect him from this hurt. You can't kiss the sore and make it go away. Simultaneously, he might feel a sense of relief and happiness that he is a much-wanted member of your family.

TAO of Hope: During this time of self-discovery and grieving, your child needs to know you will be there for him throughout his outbursts of anger, hurt, denial, and bitterness—that you will not abandon him. If, after a battle with you, he announces he is setting off to find his birth parent, don't offer to help him pack. Let him know that no question is off-limits with regard to his adoption. You might not have the answers, but you will do your best to find them. Your honesty can be tempered with compassion and wisdom. Brutal facts are not always honesty; they can be hostility in disguise. If your child's birth parents are abusive, crack-addicted young offenders, an honest answer to why they couldn't keep him can still be "They were not able to take care of you." If possible, it is best that the gory details be left to a time when your child has reached adulthood, and his own sense of self apart from others is more firmly established.

Both a fantasy and a fear that needs to be put to rest is that his birth parents will try to reclaim him. He needs to hear that they won't. Don't be lulled into believing that just because your child can speak eloquently and ask articulate questions about his adoption, he has full grasp of the meaning of the words he uses. When told that the family was "adopting" another family for the Thanksgiving holiday, eight-year-old Eliza could not understand how they could "adopt" a family if the family was not going to move in with them as she had when she was fourteen months old. The subtlety of language can cause confusion.

There needs to be a balance between being open to talking about the adoption and knowing when to be quiet if your child does not want to talk about it. Don't wait for questions to come up, but on the other hand, don't make it the only topic you discuss in depth with your child. He can take a more active role in keeping his life storybook up to date. In fact, it might be time to create volume two, with his activities in school, in extracurricular activities, and with his peers at birthday parties increasingly taking center stage and the adoption story taking a backseat for a while. He needs to feel he is more like his peers than different.

Preteens

Add to the separation/attachment struggle the beginnings of dramatic changes in your preteen's body, the desire to hang on to the past and at the same time get on with the future, stir in hormones and peer influence, and you might wish you could pack up and return to your own parents to escape this new person who shows up at the breakfast table ready to take on life and her younger brother at the same time. One moment she's all smiles, next she's in tears, and you don't have the foggiest notion why.

This same preteen, facing all the emotional, psychological, physical, and intellectual changes that her peers are facing, will add to this mix the struggle to figure out who she is separate from her family of adoption and her biological parents. Becoming more aware of her own sexuality, she grapples with a birth mother who was unable or unwilling to parent her. What was she like? What was her biological father like? Does she have siblings or half siblings? Who will she be like? How will she be different? What choices does she have in the matter? These questions, perhaps asked before, take on greater significance now.

She might struggle with a jealousy toward siblings who were not adopted, who share with her adopted parents a biological bond she does not share. In an open adoption, she might also struggle with jealousy she might feel toward any biological siblings her biological parents raised.

On the cusp of adolescence, the preteen's physical, sexual, emotional, and psychological development are not in sync. She might be able to carry on a calm, intellectual conversation about her adoption one moment and the next moment break down crying at the sight of a new mother nursing her baby, slam her biological mother for getting pregnant and not keeping her, and make you promise you will never abandon her. The three passages of grief can all be present in the same day and then not even be an issue for weeks on end.

Any answer you might have given her in regard to the "why?"s of adoption will be scrutinized. To say that "she was young and poor" doesn't cut it. "If my biological mom was too young, why do some young moms keep their babies?" "If she was too poor, will you abandon me if you become poor?" "Just because she was unmarried isn't a good enough reason to place me for adoption—other unwed moms raise their children." Your preteen needs to hear that birth parents were unable to parent her, that some children need to be adopted, and that you wanted her very much.

TAO of Hope: As the roller-coaster ride begins for your pre-teen, you might find it difficult to distinguish between what is normal behavior for any preteen and what is directly or indirectly related to the issues of adoption. Talking with parents who have raised adopted children and ones who have raised their biological children can give you a clue. If you are still concerned about any pattern of behavior, seek help from professionals who are trained in working with adopted adolescents. If there is a problem, it's a problem that needs to be attended to no matter what its roots.

Being open and honest are your best tools at this point. You are moving out of the parenting role and into the role of mentor and guide. There will be days you will be one or the other and days you will be both. Your preteen might not want either, but needs them and is counting on you to be there.

Adolescents

Take all the possible characteristics of the preteen, magnify them with the reasoning ability of an adult and the erratic emotional states of the adolescent. Add a tendency to question authority on anything and everything; throw in the critical examination of the adults in their lives, an identity crisis, and intricate peer relationships, and you have a typical teenager.

It is the identity crisis that is more complicated for adopted teens. They have two sets of parents and want to know more about their own biological, ethnic, and cultural roots. They might express anger at you for what you did or did not do to help them understand the adoption itself and all the parties involved in the adoption, accuse you of being selfish and uncaring. They might withdraw into themselves in an attempt to figure out who they really are or keep themselves very busy to avoid having to deal with these issues at all. This is the time they usually express a strong desire to find their birth parents or a strong desire never to meet "those people." It is also true that some teens are so busy with their own active and productive lives that these issues are nonstarters for them. They might approach adoption in their adult life with the emotional and intellectual maturity of an adult, bypassing any crisis over such issues during the volatile teen years.

TAO of Hope: Patience, confidence, and a willingness to weather the many storms; being present for your adolescents as a mentor and guide; and simply acting the part of the wise and caring adult is what they need most from you. Don't be defensive about their need to seek out their biological roots—it is neither an affront to your parenting ability nor an indication that they are ungrateful to you. In other words, don't take it personally. Be open to their questions, and help them find other people to talk with who have been through a similar experience.

The bonding, trust, and communication that you have nurtured over the years might appear to be vanishing. They are not gone, though they might go underground for a while as your teen struggles for independence and autonomy. This is the time to take seriously the following quote from Hodding Carter: "There are two lasting bequests we can give our children, one is roots, the other wings." You have given them strong roots; now it's time to let them test their wings, knowing they always have a place to call home.

You need to claim the events of your life to make yourself yours. When you truly possess all you have been and done, which might take some time, you are fierce with reality.

—Florida Scott Maxwell

Single-Parent Families

Yes, single-parent families are different from two-parent families. And urban families are different from rural ones, and families with six kids and a dog are different from one-child, no-pet households. But even if there is only one adult presiding at the dinner table, yours is every bit as much a real family as are the Waltons.

—Marge Kennedy, *The Single-Parent Family*

Single-parent families, also known as "lone-parent families," are created any number of ways today—some by default, some by fault, and some by conscious choice. The five most common ways are these:

1. Divorce
2. Death
3. Never marrying
4. Desertion
5. Adoption

The stigma and negative societal perceptions attached to single-parent families are slowly changing. We know that a violent two-parent family is a far worse environment for a child than a single-parent family, as is a series of foster homes, or living with drug-addicted parents. But a single-parent family is not just a better replacement for something gone bad. Nor is it simply a temporary shelter while we look for a better two-parent replacement. It is a viable kinship structure, a place where children can grow up to be responsible, resourceful, compassionate adults.

Single-parent families are formed for many different reasons. A single uncle willingly takes on the task of raising his sister's children when she becomes terminally ill. A never-married

woman reasons that an unplanned pregnancy is not a good
enough reason to get married—or perhaps it's not even an
option; her partner takes off when the reality of fatherhood hits
him. After a divorce, a mother takes a job on the other side of
the world, leaving the father to raise their children on his own.
An older, unmarried adult is approached by an adoption agency
to take on the challenge of a special-needs child. A father, partly
because of the ages and stages of his children and partly because
of the place he's at in his own life, makes a conscious choice to
remain single after the death of his spouse. The list of reasons
goes on and on.

When the issues of loss, boundary ambiguities, communication, commitment, and discipline are addressed openly and realistically, it's easier for you to see that it is not the form of your
family structure but your maturity and stability and your parenting skills that are really critical to your children's health and
well-being.

The demands and stresses on a single parent can be twice as
intense as those on parents in first-married families or stepfamilies who are living and working together to raise their children.
But a one-parent family does not have to be a one-adult family.
It's important that you not isolate yourself from friends, family,
and community groups that can offer you support, assistance,
and guidance. Children need good role models, male and
female, young and old. A grandfather, uncle, aunt, family friend,
or Big Sister or Big Brother can help children develop strong
and enduring relationships with their elders outside the immediate family circle. And your own strong circle of friends will
make it less likely that you will fall into the trap of using children as crutches, confidants, or a peer-review council. You can
be the parent, they can be the kids.

Poverty is probably one of the biggest threats to family stability. Two incomes, or one decent income, can alleviate stress in

lots of areas. In the aftermath of a divorce, many single parents find themselves below the poverty line, barely able to make ends meet. The emotional, physical, and psychological stress that this creates can make even the slightest mishap the proverbial "straw that breaks the camel's back." Whatever can be done in terms of child support, pooling of economic resources, and government assistance needs to be seen as a priority if family stability is to be maintained.

Even in the most stable of single-parent families, questions are bound to come up about the other parent or parents in this biological equation. These questions are in no way a reflection on your success or failure as a parent. They are questions that are of concern naturally to any child who has ventured into a preschool setting and seen kinship structures that are obviously different from his. (The questions and answers are similar to those addressed in the section earlier on the Adoptive Family.) In her book *Do I Have a Daddy?*, Jeanne Warren Lindsay writes a story you can share with your young children to lay a truthful and simple foundation for the inevitable questions they will ask as they grow up. You will also be faced with what to say to people who prejudge your family or make callous comments.

If and when you start dating, you will probably have mixed emotions about the whole affair: excitement about starting over, fear of being hurt again; excitement about being loved by someone, fear of being rejected; readiness to have someone to share things with and lean on, worry about how that person will fit into your life and your children's lives.

Your children will have mixed emotions as well, although the negative is more likely to come before the positive.

They might feel that you are betraying their other parent, whether he or she is dead, gone, or living down the street. If your children start to like this new person, they might vacillate between wanting to be close and not wanting anything to do with this "poor excuse for a substitute for the real thing." They might feel that they will be replaced in your life by the new person.

You might need to explain, in lots of different ways and many times, that no one can take their place in your life and in your heart. And then you have to demonstrate that what you say is really true, all the while balancing your romantic-love relationship with your parental-love relationship.

In the end, no matter how passionately in love you might be with your newfound partner, he or she must also be good for your children. It is a three-way contract this time around. Again, it is the issue of *generativity*. To make your family a safe harbor for children, the rights, needs, and wants of all parties need to be *balanced* against the final weight of what is good, just, and right for the children.

Your children might be angry that once again they are forced to make changes in their lives that they have no control over. They need to be allowed to express their feelings and know that they are listened to, cared for, and very important to you. You value their input and are conscious of using the TAO of Hope to guide you in the decisions you are making. They might not be too happy that this new person is shaking up the well-worn routines, rituals, and habits that you have all created and depended upon. A seemingly minor thing to you, like the seating arrangement at the dinner table, is not so minor to your children. The oldest child in your single-parent family might rebel at the changes in his status, duties, obligations, and privileges. Or he might indeed welcome these changes and be happy to resume the life of a teenager.

It's important that you move this new person in your life slowly into your children's lives. If she leaves and your children have become strongly attached to her, they will have one more major loss to cope with. The next time around, they might be even more cautious than they were the first time, or they might shut down emotionally to keep from getting hurt once more.

If you have worked through your own losses, taken full responsibility for your mistakes and healthy choices, and left most of your excess baggage at the last destination, you are in a good place to move forward with your children. Your own judgment and common sense are the keys to choosing wisely to stay in a single-parent family or just as wisely choosing to move from a single-parent family to a stepfamily.

Twice Upon a Time: Stepfamilies

Remarriage is an art. It requires more self-understanding than most relation-ships, as well as insight into the past that keeps an eye on the future.

—Benjamin Schlesinger

No other kinship structure has been reviled, scrutinized, attacked, stereotyped, and mythologized more than the step-family. Though sometimes called the "stepchild of the real family," structurally flawed, impaired, a poor imitation of the real thing, the opposite of an intact family, it is none of these. But, oh, does such language influence the way we as a society treat stepfamilies and the individuals in such families. To use the expression "stepchild of the real family" implies that both a stepchild and a stepfamily are less valuable, less important, less whole than their counterparts and that neither can hope to rise to the status of real. Even Julia Roberts, who plays the step-mother in the movie *Stepmom*, objected to its title because, in her words, "*Stepmom* sounds like a boogie movie. It makes me think 'horrible, evil person will slash your throat at night.' " If the person playing the character has such ugly thoughts about the word, why would anyone want to assume such a role in real life? There are lots of good reasons, none of which includes the opportunity to take on such an inauspicious title.

The word "step," in this context, comes from the Anglo-Saxon word "*stoep*," meaning bereavement or loss. All stepfamilies are born of some bereavement or loss.

Years ago most stepfamilies were formed following the death of one of the parents. Today they are formed as the result of divorce or death or two single/adoptive parents by choice.

The basic characteristics of a stepfamily are these:

- It is born of loss.
- At least one spouse is a stepparent.
- The family system is complicated and multidimensional.
- Not everyone is necessarily happy about the marriage.

A stepfamily is a real family that will often take at least three years, and sometimes more, to feel like a family linked together by a combination of love, commitment, biology, and memories. Its creation is a slow and at times methodical process, with all the ups and downs of any kinship structure. Added to the mix are myths and fairy tales, name games, the question of whether or not to adopt, teens and the potential for sabotage, sexuality in a sexually charged environment, and children of his, hers, and ours.

It is best for all parties in the combined family to take matters slowly, to use the crock pot instead of the pressure cooker, and not to aim for a perfect blend but rather to recognize the pleasures to be enjoyed in retaining some of the distinct flavors of the separate ingredients.

—Claire Berman, *Making It as a Stepparent*

Myths, Fairy Tales, and Fables

Perhaps the most important thing we can bring to a study of the family is an open mind and a willingness to accept that our own strongly held ideals of family life represent one truth but not all the truth.

—Emily Nett

Myths, by definition, are popular beliefs or traditions that have grown up around something or someone, expressing the ideals or worldview of a culture. Myths can also be nothing more than unfounded, false notions. Most of the myths surrounding stepfamilies are just that: unfounded, false notions that need to be dispelled.

- **The myth of instant love:** The expectation that a stepparent will immediately love her stepchildren because she has fallen in love with their father can put a strain on the entire family. If a stepmother believes this myth, she will feel guilty about her lack of instant love and about her ample amount of annoyance with the children. If the father believes the myth, he will be upset that his new wife is annoyed about things that he knows are just part and parcel of his children, whom he loves deeply. The children, who aren't as readily willing to buy into the myth, might in fact do lots of things to help smother any sparks of love that the stepmother might feel. There is no such thing as instant love. It takes time, patience, shared memories, and a shared history for love to develop.

- **Stepparents will love their children and their stepchildren equally:** Loving and caring are not about equality, they are about relationships, all of which are unique and not conducive to comparisons. Eliminate the need to love equally, and the possibility of even just liking a stepchild can become a realistic and workable option. You can't force love; it happens and it grows. And the fact that you come to love your stepchildren does not automatically mean they will love you in return. Love is not about payoffs and returns on investments. As Hugh Downs so aptly put it, "It is futile to love in order that we be loved in return. As soon as it is recruited to some other purpose it ceases to be love."

- **All problems are directly related to being in a stepfamily:** Not true—some are and most aren't. Most problems are related to being in a family and would be there regardless of the kinship structure— two-year-olds all say no, seven-year-olds like to assert their intellectual independence, siblings have conflicts. To frame every problem as stepfamily-related is to seek a simplified excuse for complicated interpersonal interactions. Excuses are never solutions to problems.

- **Stepfamilies will do everything the same as the family of origin:** They can't, and they shouldn't try. A stepfamily is a new kinship drama with a new cast of characters, new scenery, new script, and no director to coach the players. It is a comedy, tragedy, and mystery rolled into one. Some traditions and rituals from the family (or families) of origin can fit nicely into the new script, while others will need to be scrapped. Some scenery can be salvaged from both former homes, and new props will need to be brought in to help create a new home. Dad and his kids might want to get away together to row a canoe down the white waters of the Snake River, a yearly tradition for as long as the children can remember. Stepmom would just as soon enjoy the quiet of an empty house for a week. Mom might need time alone with her teenage daughter, while Stepdad takes all the other, much younger children on the monthly outing that has become an important part of their stepfamily tradition. The entire stepfamily enjoys an evening at the pizza parlor on Friday nights, something none of them ever did before. It's like the wedding cliché: something old, something new, something borrowed, something blue.

- **There is a best time to create a stepfamily:** There is no best time, though some times are easier than others. Creating a stepfamily with teens is not easy and can be very complicated. However, there are pluses and minuses at all ages and stages. That includes the ages and stages of the adults as well as of the children involved.

- **A stepfamily is far better than a single-parent home:** Even the reverse is not true. Staying in a single-parent home can have as many pluses and minuses as creating a stepfamily (just different pluses and minuses). You don't create one to avoid the minuses of the other, or avoid being in one to eliminate the problems inherent in it. Both work best when they are seen as viable kinship structures and not as a way to avoid the alternative.

Fairy tales are made-up stories usually designed to enshrine the stereotypes or prejudices of a particular culture. They reinforce unsubstantiated biases and create self-fulfilling prophecies. The prejudices they reinforce can be dangerous to the reputations of any well-meaning adults who are willing to take on all the responsibilities of parenting children who are not biologically connected to them, to love them unconditionally, and at times to risk their own lives for them.

The wicked stepmothers in Cinderella and Snow White would leave us to surmise that all stepmoms see their stepchildren as cheap labor. They abandon their stepchildren on a whim and would rather feed them poisoned apples than a hot supper. According to these tales, stepmoms are by nature jealous and possessive and utterly without empathy. It doesn't get much better for stepdads; Charles Dickens paints them as apathetic and uncaring. The truth is that *some* stepparents are evil, wicked, apathetic, and uncaring, and so are *some* biological parents. *Most* stepparents and biological parents are none of the above.

Children face danger in some families. We would have no need for children's-aid organizations if children were not at risk in the hands of some adults. To close our eyes to the risks in a family of origin because "parents would never do that to their own children" or "preservation of the family takes precedence over all else" is to cover our ears to children's screams for help. To immediately assume the worst when a stepparent is involved is to reinforce unsubstantiated biases and often to erroneously taint the reality of a situation.

We need fewer myths and fairy tales. What we could use are more **fables.** A fable is a narration to enforce a useful truth.

A Fable: Vindication in a Lonely Death

The people who knew Troy Tilley best never doubted that he died on a mountain trying to save the lives of two boys. But other observers—more cynical, world-weary, or just plain mean—did express doubt, which is why it became so essential that the mystery of his whereabouts be solved.

Last weekend it was, to the melancholy relief of family and friends.

And yes, they were right. The 27-year-old laundry

worker perished in precisely the manner they thought he had. He died while going to find help for his stepson, Drew Naylor, 11, and Naylor's friend, Josef Lippincott, 10, after the trio were caught in a snowstorm on Tanner Peak just south of Canon City. It was April 25, and their deadly excursion had begun as an afternoon hike.

Searchers found the boys' bodies four days later, after hypothermia had done its numbing work. But where was Tilley? He never appeared. And when day after day of searching with dogs, helicopters, and horses failed to turn up his body, people began to talk. They whispered that Tilley must have survived, but for whatever reason had chosen to vanish without telling his story. Maybe he had abandoned the boys, some said, and was too ashamed to admit it. Maybe he'd even hurt them.

One report had him spotted at a carnival in Canon City. Another placed him hitching a ride outside Pueblo. Such gossipers simply could not accept the self-evident likelihood that a good man had been overcome doing what such a man *would* do in that crisis—try to move heaven and earth in order to get the three of them to safety.

The tracker who found Tilley's body nearly a month after his disappearance said he evidently set out for help alone after making a bed and a big fire for the boys. No doubt he also tried to boost their spirits with promises that he'd be back with help before they knew it. Sadly, he couldn't deliver. Two miles away, he slipped in a small stream, fell and struck his head. "He died busting his butt trying to save those little boys," said Don Bendell.

First a man lost his way; then he lost his life. And for a while, unfortunately, he was at risk of losing his reputation, too. It's another lesson in the importance of giving people the benefit of the doubt, especially when they aren't around to defend themselves.

—**editorial in the *Rocky Mountain News***

This fable is more than a fable; it is a true story about a stepfather and how he was vindicated only in his death. The fact that he was a stepfather immediately raised doubts in the minds of many about his intentions. Rumor and mean-spirited speculations were rampant, any possibility of caring and courage on his part dismissed. As Kim Franke-Folstad expressed in another editorial after it became apparent that Mr. Tilley had indeed lost his life trying to save his stepson, "It is obviously and absolutely preferable to the version in which the wicked stepfather lures the two youngsters into the forest, then runs away and leaves them alone and frightened, their trail having disappeared." How often stepfamilies as a whole, and stepparents in particular, are doubted first and given benefit only after they are vindicated by facts.

We need to rewrite the scripts, get rid of the preconceived notions, and dismiss the myths. Only then can we realistically and without prejudice address the real concerns of stepfamilies.

The Name Game

We cannot have expression till there is something to be expressed.

—**Margaret Fuller**

At Family Night, eight-year-old Michael was attempting to introduce his entire extended family to his teacher. He started with the baby, Jamie, his new half brother; moved on to his youngest stepsister, Jill; his brother, James; older stepsister, Janice. When he got to his mom and dad and his stepmom and stepdad, he threw up his hands and said, "And the moms and dads of the whole bunch of us!"

What we call ourselves and one another speaks much more than the name itself. It tells us who we are to ourselves, to our families, to our community. We come into this world and are given a first name, perhaps a middle name, and a surname; we get nicknames, new names, and know we are in trouble when called by our full given names. We get a variation on the name when a younger brother or sister can't pronounce our full name. Sometimes we get a II or Jr. attached, a hyphenated surname, or we lay claim ourselves to an affectionate name: Nana or Gramps. Names connect us to one another and at the same time make us distinctive from everyone else. They can show our position in a family, reflect respect, show love.

Names can also present a problem. In stepfamilies, whether born of death or divorce, children come to the new family having had a biological mom and a biological dad. Now they have to find a name for this new adult in their lives. Often the adults struggle with the dilemma as much as the children. Children have been able to distinguish between Nanny Reagan and Gramma Walker for years, and no one needed to clarify that Gram was really the great-grandmother. Aunt Susie to nieces and nephews was always Aunt Susie, even after she was no longer connected by marriage to the family. Five brothers, four grandparents, any number of aunts and uncles and cousins, easy to handle—but two moms?

One big reason this is so difficult when it comes to stepparents is that the titles "Dad" and "Mom" carry with them an intimacy and closeness that can't be artificially created or easily duplicated. These names grow on you and can't be rushed. Children might fear that they will be disloyal to their dad if they call their stepdad "Dad" as well. It can also be confusing if they talk about "Dad" in a conversation with their peers. Do they mean Dad Don or Dad Patrick? The biological children might resent their stepsiblings' calling *their* dad by the title. As well, Dad might resent stepdad's being called anything that even remotely connects "that man" to his kids.

Just as it takes time for stepfamilies to function as a true family unit, the names they call one another need to evolve over time. And just as stepfamilies need to make a conscious effort to create rituals and routines and traditions, a conscious effort needs to be made to discuss the whole issue of names. The family needs to be creative in other areas of their lives, and there is no reason to be rigid in this one. There are variations on the theme of "Mom" and "Dad." One boy called his stepdad "Pops." Nicknames are an option that can overcome many of the problems associated with the value-laden "Mom" or "Dad." Nicknames come about as the result of experiences we have with one another, ways we connect and relate. They can be very personal and enduring.

As a stepparent, be patient, open, and willing to laugh as your stepchildren practice getting used to calling you something other than a pronoun. "Mom II" beats "Hey, you" or "She" hands down. And don't be surprised if you are called "Sarah" one day and "accidentally" called "Mom" the next. A forty-year-old friend calls her stepdad "John" when speaking to him and "Dad" when she is writing to him. An eight-year-old calls her stepmom "Mom" when talking to kids in the neighborhood and "Jackie" when talking to her in person. Anne and Katie were teenagers when their parents divorced and remarried. They didn't want or need another mother at that point. After practicing a few alternatives, they comfortably settled on "my dad and my Shirley."

Children need to pick a name for their stepparent that is comfortable to use and that the stepparent can live with; "extra mom" or "pretend mom" won't cut it. It's important that both parents let the children know that it's okay to practice lots of alternatives, but rude or disrespectful names are out of the question. It's just as important that a stepparent not *demand* to be called "Mom" or "Dad." One father was concerned that if his stepdaughter called him by his first name, his new son would be confused or, worse yet, not call him "Dad." His fears were unfounded. His stepdaughter called him "Dan" until her little brother began to talk. Then she and her stepdad playfully and affectionately called one another "Daughter Diane" and "Daddy Dan." Sammy called his sister "Di" and his dad "Dad." In addition he called his mom "Mom," grandmothers "Nona" and "Gram," and grandfathers "Gramps" and "Grandpa," quite capable of making all the distinctions necessary to establish with titles his own relationship to all these people.

Surnames are another concern in stepfamilies. Stepsiblings close in age have to answer a lot of questions and do a lot of explaining when they have different surnames. A mother with a new surname is easily mistaken for the stepmom, and the stepmom with the same surname as the children is just as easily mistaken for the mother—and congratulated at graduations and weddings, much to the chagrin of the biological mother. Some families have used the various surnames on legal documents and a singular surname for all other documents and activities. (Increasingly there is more than one surname in families, so this problem is not unique to stepfamilies, and we as a society are getting used to different surnames in a family, hyphenated surnames, and truncated surnames.)

Adopt or Not

Home is the place where, when you have to go there,/They have to take you in.

—Robert Frost, "The Death of the Hired Man"

Where it is an option, some families have decided that the best way to solve the problem of different surnames and create a strong stepfamily is for the stepparent to adopt the stepchildren.

This could be a solution to a lot of problems besides the surname. A young stepchild who has no contact with her biological father could well benefit from being adopted by her stepfather. If the biological parent has died, the adoption can help the child feel more rooted in the family. The ceremony and the ritual surrounding the adoption can help strengthen the bond between the stepparent and stepchild, and in the family as a whole. It is also the only way currently to create a legal bond between the stepparent and stepchild. (This is beginning to change with new laws and statutes that recognize the various kinship structures.)

There are three other reasons adoption could be beneficial for all parties concerned:

1. Medical consent: Adoptive parents have the power to make medical decisions for the child; this is not automatic for a stepparent.
2. Inheritance: A stepchild has no right of inheritance unless the proper legal steps are taken beyond a simple will. Adopted children share equal rights to inheritance with the biological sons and daughters.
3. Custody rights: If the biological parent dies, the adoptive parent will retain legal custody of the child. Stepparents, regardless of the length of the relationship or the strength of its bond, have no custody rights.

There are more reasons adoption might *not* be a good solution to what is now a stepparent/stepchild relationship:

1. Inheritance: If you adopt your stepchild, she might be relinquishing her rights to the inheritance from her biological parent.
2. Custody rights: If the marriage "goes south," you don't get to. As the adoptive parent, you are legally and

financially responsible for your adopted children, even if the children resent your involvement in their lives.

3. Money: Once you legally adopt your stepchild, all support payments from the biological parent are gone.

4. Abandonment: Perhaps the most crucial reason the adoption question needs to be thought through carefully is that with adoption comes the other side of the equation: abandonment. Unless the biological parent is dead, that parent must relinquish all rights and basically sever his or her parental relationship with the child. This might be a good thing if that is what the parent has in fact already done. However, it could also mean that the child must deal with the reality that, just as you are willing to adopt him, someone else is willing to abandon him. The emotional costs might be much higher than the many benefits of adoption.

Adoption is not a cure-all for stepparent/stepchild woes. It will not make a bad relationship better, and it could make it worse. It can also make a good relationship even more solid than before. It is one of those situations where the TAO of Hope can point the way. The adults need to be willing to compromise, to practice generativity, and to balance their rights, needs, and wants against the final weight of what is good, just, and right for the child.

Teens

[From] the adolescent's perspective, it's like discovering that another layer of management (the stepparent) is being thrust between you and the boss (parent) you've reported to for twelve, fourteen, or even sixteen years. Or worse, that the business (the home) has been bought out from under you. From the teenager's perspective, remarriage can feel like a hostile takeover.

—Laurence D. Steinberg and Ann Levine,
You and Your Adolescent: A Parent's Guide for Ages 10–20

As Laurence Steinberg and Ann Levine so wisely observe, to a teenager, the introduction of a stepparent into the family that was functioning fine without him can feel like a "hostile takeover." Another teen put it more mildly: "I felt like my stepmom arrived in the middle of a conversation, so we had to backtrack and fill her in. It was annoying; she didn't understand our inside jokes, and we missed the mom she tried to replace."

Most adults in stepfamilies and therapists who work with stepfamilies agree that the most difficult time to attempt to form a stepfamily is when teenagers are part of the mix. Teens are apt to try to undermine everyone else's hard efforts to make the family work. The three most difficult issues to deal with when it comes to teens and a stepfamily are attachment/separation struggle, history, and sexuality. The first two are straightforward; the third is complicated enough to warrant its own section in this chapter.

Attachment/Separation Struggle: One of the most important jobs for a teenager is to begin to move away from his family and establish his own identity. This means breaking away from his own biological parents, while still maintaining a sense of rootedness with them. The struggle is magnified if he is trying to break away and is being asked to join in with the new family, create new traditions, rituals, and routines together. He is also being asked to form a relationship with a new parent when he's not interested in having a parent/child relationship with anyone, let alone a stranger. Activities with peers are more satisfying than hanging with the parents, and certainly more satisfying than having a "family experience" with an adult he hardly knows and with stepsiblings who annoy him.

History: Erratic emotional states, identity crisis, questioning authority, intricate peer relationships, testing values, limits, and adults' patience are hallmarks of a teen. The biological parent can "remember when" he was an easy-to-get-along-with, happy-go-lucky, friendly, considerate ten-year-old. The stepparent sees only this defiant, mouthy, irresponsible, pushy kid who bears no resemblance to the picture of the smiling child on the family mantel.

Still think it's worth it? It is. There are a lot of reasons why a solid stepfamily is good for teens. Three important ones are:

1. Stepfamilies can be healthier than the families of the first marriages, as long as all the adults involved are willing to learn from the problems in the first marriage, leave most of their excess baggage behind, and make a concerted effort to compromise, embrace generativity, and create a safe harbor by balancing the rights, needs, and wants of all parties against the final weight of what is good, just, and right for the kids.

2. When parents are happy, it benefits teens. Spousal conflict in first marriages can do a lot of harm to kids (see "Peaceful Resolution to a Violent Marriage" in the previous chapter). Being free of that conflict allows kids to spend energy on handling the normal conflicts in their own lives. A stepfamily can be a place to get a fresh start and create new memories. It can also be a role model for a good marriage and instrumental in helping teens develop the ability to form good love relationships.

3. As long as a stepparent is willing to be present but not overbearing, to be a mentor and a guide, that teen will have one more positive role model and one more supportive, caring person in her life. It's impossible to have too many loving people in your life.

Sexuality—Yours and Theirs

A major developmental task for families at all stages of the life cycle focuses on sexuality. . . . [A] sexually healthy family is not only one that avoids the extremes of sexual neglect and sexual abuse. It also actively promotes the sexual well-being of all its members through unique and integrated patterns of family interactions.

—Phyllis Meiklejohn et al., *Today's Family: A Critical Focus*

Put newlyweds together with a bunch of kids of all different ages and stages of sexual development, blur the boundary lines of relationships, add communication problems, commitment challenges, and discipline, and you have a stepfamily facing the challenge of creating a sexually healthy family in a sexually charged environment.

Intimacy and sexuality are different in a stepfamily than in a first marriage. In this second time around, you often have an audience of children who are curious about their own sexuality and embarrassed by the openly affectionate display of yours. Your teens are embarrassed also about the changes in their own bodies and would really rather you not behave the way they want to.

Stepchildren might be attracted to the stepparent, the step-parent to a stepteen, and stepsiblings to one another. Usually in the biologically connected family, the boundaries between affection and sexuality are clearly defined; the children and parents have a history together of physical touching that is natural and innocent. Physical affection, bantering, and jest are a normal and welcomed part of their relationship and the relationship with siblings. This history is not there in newly formed stepfamilies.

The sexual and emotional maturity of the adults is paramount if the children are to get the help they need in respecting the healthy boundaries of romance and sexuality. Both adults must be committed to promoting the sexual well-being of everyone in the family. Open eyes and open communication are vital. When anyone in the family expresses an uncomfortableness, listen and be aware. The uncomfortableness could be a red flag that something is amiss.

When preteens or teens are part of the stepfamily mix, it helps if some simple ground rules are established *before* the two families come to live together, but any time is better than never. Some of the issues to be covered: bedroom and bathroom etiquette (knock and lock); what is appropriate to wear around the house (scant is out, robes are in); weekends without adult supervision (not even in the cards); continuing dialogue on sexual issues (open for discussion anytime).

When a stepteen is attracted to his or her stepparent, it is the stepparent who must demonstrate self-discipline and give guidance concerning what comments and behaviors are appropriate and healthy.

Stepsiblings who are thrown together in a family constellation not of their own choosing, living in close quarters with unclear boundaries, new routines, and lots of time together might find themselves sexually attracted to one another. The incest taboo is not applicable here; as far as they're concerned, they're not related. This is where the complexities of a stepfamily come into play once again.

Sexual attraction will not necessarily diminish the ability of the family to function as a unit, but sexual involvement will. The adults in the family can help the teens to understand the difference between feelings, fantasies, and behaviors. Denial on the part of the parents ("Oh, they are not serious" or "It's just puppy love; they'll get over it") hurts everyone involved. Demanding that the relationship cease can move it underground and as a result damage any healthy communication between the teens themselves and between you and them. By bringing it out into the open and talking about the feelings, you can help your teens find ways to "evolve" these feelings into a friendship.

It is also possible that once the teens leave home, they might find that not only do they have a strong friendship, they are truly drawn to one another. As adults, they might date one another and perhaps fall in love. It's the possibility of now "falling out of love" that can prove difficult and complicated. Since they must meet one another at family holidays and celebrations, they will need to make the transition from lover to stepsibling with the knowledge that once some boundary lines are crossed over, there is no way just to go back to what was. Their relationship is forever changed—sometimes for the better, sometimes for the worse, but definitely changed.

Be aware that stony silence or stormy confrontations might be cover-ups for sexual feelings your teens are embarrassed or shamed by. You can bring up with your teen the subject of sexual attraction. If, in the end, it is not sexual attraction being covered up, but disgust, anger, or outright dislike of the sibling, at least you've opened the door for further communication on sexuality and at the same time helped your teen deal with all those other feelings.

If there is any risk for abuse, or if the family is breaking apart emotionally because of an overly sexually charged environment, get help fast. It is much easier to prevent sexual problems from growing into a family crisis than it is to try to repair the damage once the lines have been crossed and boundaries violated.

"Healthy" family sexuality can be generally defined as: the balanced expression of sexuality in the structures and functions of the family, in ways that enhance the personal identities and sexual health of individual members and the coherence of the family as a system.

—James W. Maddock, "Healthy Family Sexuality: Positive Principles for Educators and Clinicians"

His, Hers, Ours

Don't decide whether or not to have a child by its possible effects on the stepchildren. Decide it the way you would if you were a husband and wife with no stepchildren to consider. That is, if having a child is important to you and your marriage, then have it. If not, don't.

—**Dr. Fitzhugh Dodson**

Dreams are dashed, hopes are buried, positions of status are surrendered, jealousies resurface, relationships are forever altered—and we're talking here about the birth of a child into a family, a child wanted by at least two people in that family and the only one related to everyone else in it. When a child is born into a stepfamily, it is a monumental event that creates monumental changes for all the other children. As excited as you are to be the proud biological parent again or for the first time, your children and/or stepchildren might not immediately share in your excitement and bliss.

For starters, the fantasies or hopes of the children that their respective biological parents will get back together are destroyed. Up until this time, there was always the remote possibility that Mom and Dad might divorce their present spouses and remarry one another. Second, an "only" or a "last-born" is relegated to playing second fiddle to this new youngest and only biologically-connected-to-everyone baby. Third, questions abound: "Will Mom and Dad still love me?" "How am I related to this baby?" "When I live with my mom next week, does the baby come along?" "Do I have to share my room again?" "Why did they have her? Weren't we enough?" "Are you going to have more?" And as if that weren't sufficient stress, older siblings are likely to throw in a few caustic comments: "It's disgusting to have another child at your age." "Don't ask *me* to baby-sit that baby." "You never have time for me as it is; I'm going to live with Mom. At least she acts her age." "Hope you do a better job of parenting this time."

After reading all this bad news about having a child together, are there any good reasons even to consider the option? Lots, and none of them good enough in and of themselves to be a good enough reason. As Blaise Pascal said, "The heart has reasons which reason knows nothing of." It must be for a heart reason that you choose to have a child together. You can factor all the pluses and minuses and still not come up with the right answer. Listen to children's free advice on the matter, but don't ask them for permission.

There are pluses to having this child together. As parents, you have a chance to reevaluate your attitude and behavior toward all the children. Becoming a biological parent for the first time can help make you an even better stepparent. The rituals and traditions surrounding the birth of this child can help create stronger family ties and boundaries. It's a joyous moment when you hear your other children announce to their peers, "*We* have a new baby!" Even the jealousy the stepsiblings feel can help them draw closer to one another as they try to make peace with this new enemy in the camp.

A child is like a precious stone, but also a heavy burden.

—Swahili proverb

Being born of both loss and hope, there are things that successful stepfamilies can teach all of us: Of necessity they are masters of negotiation and compromise. They can balance boundary ambiguity with family ties. Being flexible, they are comfortable alone, with a few family members, and with many family members. They tend not to be exclusionary. And they have an intimate understanding that loss does not mean the end of the world. People can pick up, move on, and be strong.

Looking back we see with great clarity, and what once appeared as difficulties now reveal themselves as blessings.

—Dan Millman

Effective Child-rearing

What appears to be crucial to effective childrearing is not so much the particular kinship structure as the emotional climate of the family. Authoritative parents who set firm limits with love and thoughtfulness, are more effective than strict, authoritarian parents or permissive, laissez-faire parents. And this is true regardless of the particular kinship structure of the family.

—David Elkind, *Ties That Stress*

Effective child-rearing is one of the primary goals and one of the most perplexing problems for any family. The key is to believe that you are a family first, with some unique aspects as the result of being a stepfamily, single-parent family, or adoptive family. The following is the description I wrote in *kids are worth it!* of a "backbone family" that is created by the authoritative parent. By concentrating on creating a healthy emotional climate at home—in the face of adversity, chaos, loss, sibling rivalry, financial struggles, another trip to the dentist, a call from your former spouse's lawyer, and inevitable conflict with your present spouse—you will find that you have the resolve it takes to make your particular kinship structure a truly effective family.

Backbone families . . . are characterized not so much by what they do or don't do but by how they balance the sense of self and the sense of community in all that they do. Interdependence is celebrated.

Backbone families can also be described by what they are not: They are not hierarchical, bureaucratic, or violent. Backbone parents don't *demand* respect—they demonstrate and teach it. Children learn to question and challenge authority that is not life-giving. They learn that they can say no, that they can listen and be listened to, that they can be respectful and be respected themselves. Children of backbone families are taught empathy and love for them-

selves and others. By being treated with compassion themselves they learn to be compassionate toward others, to recognize others' suffering, and to be willing to help relieve it.
The backbone family provides the consistency, firmness,
and fairness as well as the calm and peaceful structure
needed for children to flesh out their own sense of true
self. Rather then being subjected to power expressed as
control and growing up to control others, children are
empowered and grow up to pass what they have learned of
the potential of the human spirit on to others. . . . Backbone families help children develop inner discipline, and
even in the face of adversity and peer pressure, they retain
faith in themselves and their own potential.

A new family can be a new beginning for all members. No
one has to keep doing what they have always done, or do what
was done to them. The key is to recognize the messages and
tools you received from your own parents and are still carrying
around, become conscious of the messages you are giving your
children directly or indirectly, and become aware of the emotional and physical environment you are creating as a family.

*The terrible truth I see, in all the research devoted to this book, is that good
people create good families, and vice versa. This is not a soothing conclu-
sion, because we don't know how to create virtue. But from classical Rome to
Confucian China to Los Angeles today, the structure of the family seems to
matter less than the ways that we choose to behave within it.*

—Suanne Kelman,
All in the Family: A Cultural History of Family Life

Part Four

· · · · · · · · · · · · · · · · · · · ·

RESPONDING TO CRISES
LARGE AND SMALL

Chapter 6

Mistakes, Mischief, and Mayhem

Do justice . . . love mercy . . . walk humbly . . .

—Micah 6:8

In a moment of exuberance, your four-year-old moves a ballpoint pen past the edge of her drawing tablet and scratches the wooden kitchen table. In a fit of boredom, your nine-year-old uses the same pen to gouge several permanent tic-tac-toe patterns into the same table. Returning from a weekend trip, you stand at the kitchen door in shock as you survey the damage to the entire kitchen, table included. Your teenager tries to explain that she had a few friends over and the party got a bit out of hand. The most serious damage was done by your daughter's best friend, Sam. Fueled with alcohol, and angry and bitter about his parents' acrimonious divorce and ongoing custody battles, he carved a variety of four-letter words into the table.

What's a parent to do? We might be inclined to let the mistake slide, punish the nine-year-old by removing his Game Boy for a week, ground our teenager for six months, and ban her friend from our property forever. These "solutions" are a variation of punishment or its alter ego, rescuing. None disciplines the kids or helps them to develop their own sense of inner discipline. And for the angry teenager, there is no opportunity to fix what he did, figure out how he can keep it from happening again, or heal with the people he has harmed.

How we respond to their many mistakes, occasional mischief, and rare mayhem can help provide the wherewithal for our children to become responsible, resourceful, resilient, compassionate humans, who feel empowered to act with integrity and a strong sense of self, or to become masters of excuses, blaming, and denial, who feel powerless, manipulated, and out of control. Whether they feel empowered or powerless will greatly influence their ability to handle the myriad traumas they will experience throughout their lives, traumas brought about through death, divorce, illness, natural disaster, broken friendships, loss of a job, or mistakes, mischief, and mayhem they create themselves.

Moral autonomy appears when the mind regards as necessary an ideal that is independent of all external pressure.
—Jean Piaget, *The Moral Judgement of the Child*

Discipline and Punishment: Why One Works and the Other Only Appears To

If you punish a child for being naughty, and reward him for being good, he will do right merely for the sake of the reward; and when he goes out into the world and finds that goodness is not always rewarded, nor wickedness always punished, he will grow into a man who only thinks about how he may get on in the world, and does right or wrong according as he finds of advantage to himself.
—Immanuel Kant, *Education*

Although the words are often used interchangeably, discipline is not synonymous with punishment. Punishment is adult-oriented, imposes power from without, arouses anger and resentment, and invites more conflict. It exacerbates wounds rather than heals them. It is preoccupied with blame and pain. It does not consider reasons or look for solutions. It is doing something *to* a child when the child behaves in a way that the parent judges to be inappropriate or irresponsible. It involves a strong element of judgment and demonstrates the parent's ability to control a child.

Punishment preempts more constructive ways of relating to a child. It drives people further apart, and it enables the parent and child to avoid dealing with the underlying causes of the conflict. The overriding concerns of punishment are these: What rule was broken? Who did it? and What kind of punishment does the child deserve? Punishment discourages the child from acknowledging her actions ("Wasn't me, didn't do it"). It deprives the child of the opportunity to understand the consequences of her actions, to fix what she has done, or to empathize with the people she might have harmed. It increases tension in the home, and it helps children develop a right/wrong, good/bad distorted view of reality. "Good behavior" is bought at a terrible cost.

Punishment leaves control in the hands of the parents (sometimes literally) and gives children the message "I, as an adult, can and will make you mind," often with the rationale "for your own good." Its goal is instant obedience. Hitting a child for every mark he made on the table doesn't teach him not to destroy property; it can teach him to avoid getting caught, and it can teach him that "might makes right." It becomes a tool he himself can use when his brother won't give him back his toy. Aggression begets more aggression.

More often than not, the tools used to attempt to control a child and make him "feel the pain" for what he has done are subtler than physical force. They can take the form of any of the following:

Isolation: "Sam, you are never to set foot in this house again."

Embarrassment and humiliation: "You will have to eat your dinner on the floor for the rest of the week; any child who writes on a table doesn't deserve to eat at it."

Shaming: "If you are going to act like a three-year-old who can't be trusted when I leave, I am going to treat you like a three-year-old."

Emotional isolation: "Stay out of my sight."

Grounding: "You are grounded for six months, Senior Prom included!"

Brute force: "You get one hit for every mark you made on the table."

Illogical consequences: "I'm taking your Game Boy away for a week. Maybe then you'll show some respect for the furniture."

With these forms of punishment there is only an arbitrary connection between what the child has done and the resultant punishment. It can be a stretch for the child to try to figure out how the deed and the punishment go together. Reason suffers. As well, all of these tools can degrade, humiliate, and dehumanize children. Embarrassment, humiliation, and shaming might make a parent feel good, but they're unlikely to change the behavior of the targeted child. He will probably want to hide and likely avoid taking responsibility for wrongdoing, concentrating more on how badly he's being treated than on what he did that initiated the punishment.

The mind-set of the parent is that a rule has been broken and punishment must be imposed. Under the guise of discipline, physical and emotional violence are legitimized and sanctioned. Children might behave so as not to get caught, but their sense of self-worth, their sense of responsibility, and their sense of appropriate, responsible, caring actions are seriously compromised. They often respond to punishment with the three F's: Fear (doing as told out of dependency and fright), Fight (attacking the adult or taking the anger out on others), Flight (running away mentally, afraid to make a mistake or take a risk, or running away physically).

In *Living Faith*, former president Jimmy Carter speaks to the problem of relying on punishment and the resultant fear of retribution to teach children to do good:

> There are many unenforceable standards in our private lives. . . . If we are interested in lives that excel, we will wish to do more than just obey the law. How do we act when there is no accountability for what we do? What restrains us from being rude to others, ignoring the plight of needy people, giving false information when it is to our advantage, abusing a defenseless person, promulgating damaging gossip, holding a grudge, or failing to be reconciled after an argument? These are the things for which we will not be punished, and therefore the fear of retribution is missing as a motivation.

With punishment, children are robbed of the opportunity to develop their own inner discipline, the ability to act with integrity, wisdom, compassion, and mercy when there is no external accountability for what they do.

Discipline, on the other hand, is not judgmental, arbitrary, confusing, or coercive. It is not something we *do to* children. It is a process that gives life to learning; it is restorative and invites reconciliation. Its goal is to instruct, teach, guide, and help children develop self-discipline—an ordering of the self from the inside, not imposition from the outside. In disciplining our children, we are concerned not with mere compliance but with inviting our children to delve deeply into themselves and reach beyond what is required or expected.

The process of discipline does four things the act of punishment cannot do:

1. It shows kids what they have done.
2. It gives them as much ownership of the problem as they are able to handle.
3. It gives them options for solving the problem.
4. It leaves their dignity intact.

For mistakes, mischief, or mayhem that intentionally or unintentionally create serious problems of great consequence, the three R's of reconciliation are incorporated into the four steps of discipline. These three R's—Restitution, Resolution, and Reconciliation—provide the tools necessary to begin the healing process when serious material or personal harm has occurred. Whether it is only the four steps or the four steps and the three R's, discipline deals with the reality of the situation, not with the power and control of the adult. It helps change attitudes and habits that might have led to the conflict, and it promotes genuine peace in the home.

Discipline involves intervention to keep a child from further harming himself or others, or real-world consequences, or a combination of the two. Real-world consequences either happen naturally or are reasonable consequences that are intrinsically related to the child's action. (To refinish or replace the tabletop that had four-letter words carved into it would be a start in the reconciliation process.) Real-world consequences take a bit of reasoning but not a lot of energy on the parent's part, and certainly they shouldn't be a struggle. Discipline by its nature requires more energy on the part of the child than on the part of the adult. If a consequence is RSVP—**R**easonable, **S**imple, **V**aluable, and **P**ractical—it will invite responsible actions from the child. If it isn't all four of these, it probably won't be effective, and it could be punishment disguised as a reasonable consequence.

Often such disguised punishment is predetermined and is based on the assumption that all violations are clear-cut. Carving the table would be a violation subject to a one-size-fits-all punishment, regardless of the intent of the violator. In our rush to swift and certain judgment, there is no place for discernment of intent; the deed is seen only as a violation of a rule: Children do not carve in tabletops. Even a mistake unpunished is looked upon as a possible misstep down the slippery slope to more violent deeds. ("If she gets away with these marks on this table, she'll probably think she can get away with carving up the good dining-room table. I have to do something to her, or she'll think she's done nothing wrong.") Such a mentality of zero tolerance creates an environment of zero options for parents. It is a simplistic response to complicated actions. It wrongly presumes that a young offender created the mayhem with the foresight, judgment, and maturity of an adult.

The opposite extreme (punishment's alter ego) is rescuing a child because we believe that children are incapable of wrongdoing with malevolent intent. We make light of the incident, ignore it entirely, or devise excuses for the behavior. If we don't draw attention to it, maybe it will just go away. This is just as wrongheaded as the punitive approach. Overcome by the sympathy we feel for the perpetrator, we try to convince ourselves that if we only knew the reason for the child's misdeeds and the history that preceded the mischief or mayhem, we would be compelled to forgive and forget. Punishment ignores intent; rescuing ignores the severity of the deed.

Discipline is a more constructive and compassionate response that takes into consideration the intent, the severity of the deed, and the restorative steps needed to give life to the child's learning and to heal relationships that might have been harmed. It invites us to respond to our children with mindfulness, reason, a wise heart, compassion, and mercy, instead of just reacting with logic or emotion. It enables all of us to go beyond mere repair to restitution, resolution, and reconciliation.

During times of chaos and loss, it is children who have experienced such discipline, instead of punishment, who will have an inner reserve or resource to draw on when their strength is sapped, their intellect assaulted by the answerless questions, their emotions thrown into turmoil by raw, piercing grief.

Building a conscience is what discipline is all about. The goal is for a young-ster to end up believing in decency, and acting—whether anyone is watching or not—in helpful and kind ways.

—James L. Hymes, Jr., "A Sensible Approach to Discipline"

Mistakes

Learn from your mistakes is an old rule, but it is surprising how many people fail to heed it . . . [I]f you can't see the lesson in what went wrong, you're just condemning yourself to make the same mistake again.

—Chuck Norris,
The Secret Power from Within: Zen Solutions to Real Problems

The four-year-old who has scratched the table cannot repair it. The reality is that the table has been scratched. The problems to be solved are (1) how to minimize the mark, and (2) how to keep the table free of further "accidental" marks. The pre-schooler can wipe the table to remove the color from the mark-ing and help place a larger mat under her drawing paper to protect the table from any damage the next time she is drawing. You can give her felt-tipped pens—easier to clean up and less likely to cause irreparable damage. As a parent, you are there to give your child as much ownership as she is capable of assuming for the problem she created, offer guidelines for fixing the mis-take, and assure her that she can handle it.

To remove all pens from her and never let her draw again at the table is punishment that teaches her nothing about learning from her mistakes. She also learns nothing about being capable of fixing mistakes she has made or about how she can keep from making the same mistake in the future. To spank her for being careless is to invite her to become fearful of ever making another mistake, to hide her mistakes, to strike back at you if she dares, or in anger and hurt go hit her younger brother or the cat. To let her continue to mark up the table, even accidentally, is to say that she need not be concerned about any limits and boundaries. To offer excuses—"She's too young to know any better" or "She didn't mean to do it" or "All kids scribble on tables"—is to teach her to make excuses for her future mistakes: "It wasn't my fault." "She made me do it." "I couldn't help it."

Everyday incidents and mistakes can be opportunities for children to take ownership of problems they have created, figure out how to fix the problems, and recognize how to keep the same mistake from happening again. Kids, even at age four, begin to see that when they have a problem, what they need is a good plan, not a good cover-up and not a good excuse. As they grow older, they will be less likely to dread taking risks that might result in great failure (or great success). Rather than giving up when they experience setbacks and defeats, they can be open to learning from adversity and using that knowledge to create new opportunities.

Mischief

Do not find fault, find a remedy.

—Henry Ford

The four-year-old did not intend to damage the table; the nine-year-old did. He needs to go through all four steps of discipline, with special attention to how he is going to keep such damage from happening again. Taking ownership of the problem he created, he will need to take part, to the extent that he is capable, in getting the table repaired. He can help sand and refinish the tabletop if you, the parent, have those skills and can help him learn how. If not, he can make a phone call to the refinisher, help deliver the table to the shop, help carry it back into the house, and arrange a reasonable payment plan to you for the repair. (If the repair costs are way beyond the means of repayment for your child, you can show some mercy and reduce the debt to a workable amount, covering the excess yourself. You would hope your in-laws would do likewise if you were to break a piece of their cherished and very expensive crystal.)

All of this will involve your time as well as your child's time. Punishment is so much swifter; doing it yourself so much smoother. However, the time you take is well worth it as your child begins to realize that all his actions have consequences. He also learns that he is quite capable of taking ownership for what he does, and just as capable of taking full responsibility for the problems he's created, not because he fears reprisal but because it is the healthy thing to do.

Mayhem

Though we want to believe that violence is a matter of cause and effect, it is actually a process, a chain in which the violent outcome is only one link.
                              —Gavin de Becker, *The Gift of Fear*

Standing in the kitchen feeling betrayed, wronged, hurt, disappointed, and angry, you know that discipline is only the first step in restoring both the table and the relationship you had with Sam, your daughter's friend who destroyed this family heirloom. How easy it would be to display righteous indignation, continue being angry and feeling victimized, to push for punishment, seek revenge, and hold a grudge. Just as easy and as unproductive would be to make excuses for the teen or shrug your shoulders and chalk the experience up to a once-in-a-lifetime outburst that won't happen again and need not be addressed further.

There is a real need for the teen to take ownership of the mayhem he created, fix what he did, figure out how he can keep it from happening again, and heal with the people he has harmed. This cannot happen in an atmosphere of punishment, vindication, or vengeance; nor will it happen in an atmosphere of indifference. It can happen only if we are willing to create an atmosphere of compassion, kindness, gentleness, and patience, in which we can help him work through the four steps of discipline and the three R's: restitution, resolution, and reconciliation.

Restitution

Wise people seek solutions; the ignorant only cast blame.

—TAO 79

The first R, restitution, means fixing what he did. It involves fixing both the physical damage and the personal damage. The table might be easier to repair than the personal rift created by the act of damaging it. Fixing the table or paying for its replacement is usually less painful than the act of true repentance. And it is only such repentance that can move the offender toward reconciliation with those he has harmed by his deed. True repentance makes no room for excuses ("I was drunk"), blame-shifting ("They dared me to do it, and they knew I was drunk"), "but"s ("But the table had marks already before I carved on it"), "if only"s ("If only you hadn't left for the weekend. If only my parents weren't getting a divorce"). Repentance is not the obligatory "I'm sorry" that is used to express regret or remorse when one is caught doing something wrong. To repent honestly and unconditionally means to lament the damage caused, not out of a sense of duty or of obligation but out of a heartfelt need to admit the wrongness of what's been done, to express a strong desire not to do it again, to assume responsibility for the damage, and to begin to mend the torn relationship.

You cannot force repentance on someone else. You can help Sam arrive at repentance by helping him work his own way through the three R's. Repentance is not a goal in itself. Rather it is a by-product. It comes about only as Sam works through the whole process of reconciliation. As a wise and caring adult who is not out to rescue or punish Sam, you can provide the structure, the support, and the permission he needs to begin the process.

Resolution

. . . there is no mystery of human behavior that cannot be solved inside your head or your heart.

—Gavin de Becker, *The Gift of Fear*

The second R, resolution, means figuring out a way to keep it from happening again. In other words, how can Sam create himself anew—not apart from what he has done, or in spite of what he has done, or because of what he has done? Creating anew involves integrating the past destructive act and all its results and implications into a new beginning. It happened. He can't go back and wish it not so. He needs to be able to figure out what he actually did, what he did to bring it about, what he can learn from it, what he can take from the experience to, as Ernest Hemingway described it, "become stronger in the broken places." Without such resolution, the repentance becomes a hollow regret, a mere apology to be repeated when he's caught destroying property in his next drunken rage. True repentance requires that he redirect his destructive energies in more constructive ways. It's not enough simply to say it won't happen again. Sam needs a plan and a commitment to make that plan happen.

For his plan to be effective, it must be much bigger than figuring out a way to replace the table. His plan might need to include getting help for his drinking problem, figuring out constructive ways to express the hurt and anger he feels about his parents' constant battles, and developing a positive game plan in advance of the next party. By showing your support of Sam and his plan, you can also open the door to a frank discussion with his parents about what you have seen firsthand in terms of their acrimonious divorce and its effect on their son.

Reconciliation

We have been called to heal wounds, to unite what has fallen apart, and to bring home those who have lost their way.

—St. Francis of Assisi

The third R, reconciliation, is a process of healing with the people you have harmed. It involves a commitment by the offender to honor his plan to make restitution and live up to his resolutions. It also involves a willingness on the part of the person offended to trust, to risk, and to rebuild a relationship with the offender.

It is helpful if the offender, after making restitution, offers his time and talents to those he has harmed. This serves two purposes: One, the person harmed can experience the inherent goodness of the offender; and two, the offender can experience his own inherent goodness.

Most young offenders would not come up with this step on their own. They would like just to stop at step two and be done with the whole ordeal. It is the adult in the equation who needs to push for this step, as much for herself as for the young offender.

It will take time for you to be open to reconciling with Sam. The first night you might be in shock, hoping to wake up in the morning to find that the scene in the kitchen was just a nightmare. But in the morning, you realize it is not a bad dream.

Your thoughts turn to revenge and your feelings to anger, or to loss and sadness, or a combination of all four. If you will be honest and forthright with your thoughts and feelings you'll begin to see clearly what you are angry or sad about and what you need from Sam. Is it the wanton destruction, or the fact that the table was a family heirloom, or the words that were written, or a combination of all three that upsets you? Can the table be repaired, or does it need to be replaced?

Even as you go through the motions of hearing him out with his plan to fix what he did and his resolutions, you might find yourself grieving over the loss of the table, the loss of trust, and the rift created by the losses. You can't just forget the incident and get on with your life as if it never happened.

To try to cover up your grief will serve neither of you well. To make light of the loss ("It was only a table. It could have been worse. We can always get another one") will help you absorb the loss into yourself—and it will stay there to fester. To shrug off your own feelings ("I'm not as angry about the table as I am concerned about how you're doing, Sam") is caring, but for only one party in the reconciliation. When your caring is unidirectional, it says your feelings don't count, and uncounted feelings can turn into depression. An eighteenth-century monk spoke of such unbalanced caring: "Living the truth in your heart without compromise brings kindness into the world. Attempts at kindness that compromise your heart cause only sadness."

If you will give yourself time to move through the anger and honor your sadness, you will find yourself ready to look for creative solutions to solving the problems that both you and Sam face in order to be reconciled.

Time, in and of itself, does not heal relationships, but it does take time to heal. Even if Sam comes to you the next day with a heartfelt apology and an offer of restitution, you might need to ask for a bit more time before the two of you can truly reconcile. The *intention* behind asking for time is not to hurt him or make him suffer longer for what he did. It is time to face hurts, vent your emotions, and begin to release any grudges and destructive feelings so that you not only reclaim your own peace of mind, you open your heart and your hands to reconciliation with Sam.

To wallow in your feelings, be they anger or sadness, is to deny yourself the opportunity of reconciling with Sam. It also locks you into viewing yourself as a helpless victim and Sam as an oppressor. You could find yourself beginning to divide your world into victims and oppressors, us and them, separate from one another and unequal. You could spend your days vengefully scheming ways to punish Sam, or spend those same days weaving your garb of victimhood, declaring that your trust has been irrevocably broken, that anything Sam does will be too little too late. Your demands for restitution and resolution become vengeful and next to impossible. Either way, you end up stressed out, isolated, and possibly in conflict with everyone around you. The table, family heirloom that it might have been, becomes larger than life. And your life becomes miserable. Far better to acknowledge your thoughts and your feelings and begin to work through them to arrive at a place where you are truly ready to commit to the reconciliatory process. Along the way you will become freer, not unmindful of what has happened but unchained to the event.

Fear grows out of the things we think; it lives in our minds. Compassion grows out of the things we are, and lives in our hearts.

—Barbara Garrison

Quiet Time and the Teddy Bear

Our children need help learning how to play, how to create their own routines and what to do when they feel angry, lonely, or bored.

—Charles Schwarzbeck

The three R's are not just for older children. When a child as young as four bops her sister over the head because "she wouldn't give me the teddy bear, and I asked her for it nicely," it is time to move beyond the basic four steps of discipline and teach the hitter how she can fix what she did, figure out how to keep it from happening again, and heal with the sister she has harmed.

A parent might be inclined to dismiss the hit as mere childish antics and make excuses for her daughter ("She didn't really mean to hit her; she just gets a little excited sometimes"), or plead ("Please be nice to your sister. She's smaller than you"), or in desperation buy two teddy bears just to "keep peace in the house."

A parent might be inclined to hit the girl and scream ("Don't you ever hit your sister again"), threaten ("If you hit her again, I'll take all your toys away"), shame her ("You are a bad girl. I don't like girls who hit"), or send her to her room for a "time-out," count the number of minutes she is to stay there, and make her say "I'm sorry" to her sister. If she refuses to go to her room, the countdown begins: one, two, two and a half, two and three-quarters. She doesn't move. You grab her and forcefully put her in her room and spend the next five minutes playing doorknob-pull. Finally she calms down, you think. No, she's throwing toys out the window, so the entire neighborhood knows you have timed your kid out.

If the child has figured out the routine, she might hit her sister and go sit in the time-out chair before you even send her, knowing she need only sit the required time, say "I'm sorry," and hit her sister again when it's advantageous to do so, repeating the routine of hit, time-out, "I'm sorry," and hit again.

Neither rescuing nor punishing the child will serve her well. Discipline will. You are quick to respond: "You're angry. It's okay to be angry. It is not okay to hit. You need time to calm down. You can calm down in your room, in the rocker, or on my lap—take your pick." (Notice, there are three options. You give two, and a strong-willed child will try to figure out which one you want her to pick and purposely choose the other. Give three, and she'll be confused.) A really strong-willed child might even declare that she's not moving and you can't make her. Wisdom just went out the window; all you have left is wit: "That's a good place to calm down, too. I hadn't thought of that one." It's really hard for a child to stay angry when she's given permission to stay where she just said she was going to stay. There is no battle for control. Your goal was not to make her go to the bedroom. It was to get her to calm down so she could go on to the next step.

Once she has calmed down, she needs to fix what she did. If in anger she threw the toy across the room, now is the time to pick it up, not before when she was still angry. An apology is in order, but it is *requested*, not demanded. If you demand an apology you will get one of two kinds: an insincere "I'm sorry," or the hit/obligatory "I'm sorry"/hit again cycle. An apology is more likely to be forthcoming if the child has seen it modeled or has been the recipient of a sincere "I'm sorry."

The second step is resolution: She needs to figure out how she can keep it from happening again, which doesn't mean she can merely say, "I won't hit her again." That's what she *won't* do. She needs to know what she *will* do when she wants the teddy bear and her sister doesn't want to give it to her. This is where your wisdom and teaching come in. You might have to leave the chores for later and take the time to teach some basic skills of communicating and playing together. Most children's TV shows, mass media, and video games are no help here.

You can teach them both that if one does not want to give up the teddy bear, the other must accept the fact that the tools she has used cannot make her sister give the toy to her. A no needs to be honored as such, and she will have to go find another toy to play with for now. This is a difficult thing to accept at four, but what an important insight to learn and begin to understand. That one can say no and mean it and another can honor that no will serve both of them well in the teen years.

The third step, reconciliation, means to heal with the sister she has harmed. She can find a way to help her sister have a better day, and share in the adventure. Getting hit over the head was not a good start. The one who hit might offer to pull her sister in the wagon, since she knows her sister likes that. The younger sister experiences the goodness of her older sister, and the older sister experiences the same goodness. They are now ready to go on to play, and probably even on to new altercations that day, but with a clean start for both.

This step is far easier for a four-year-old than for a forty-year-old. You are at a family gathering, and you've just verbally ripped your brother apart with lines you've been saving for weeks in retaliation for the lines he zinged at you at the previous family gathering. Berating your brother is easy. It's meeting him at the next family gathering that will be difficult if there hasn't been a reconciliation first. Some siblings never get back together, refusing for years to sit next to one another, not remembering exactly why, but knowing that they have "good cause" for keeping the grudges and resentments alive. Both are trapped in their anger, neither willing to give, unable to extend mindfulness and compassion to one another.

If self-discipline and the three R's can become a part of your children's everyday encounters with their siblings, their peers, and the adults in their lives, they will be able to replace the destructive tools of revenge and retribution that were your well-worn family heirlooms, passed on from generation to generation, with lighter, more constructive tools.

It's not just this or another action which makes us peacemakers, but it is the entire fabric of our lives.

—Mahatma Gandhi

Nonviolent Engagement and Reconciliatory Justice

Nonviolence is a flop. The only bigger flop is violence.

—Joan Baez, *Daybreak*

Nonviolence can never be equated with passivity; it is the essence of courage, creativity and action. Nonviolence does, however, require patience: a passionate endurance and commitment to seek justice and truth no matter the cost.

—Mary Lou Kownacki,
Love Beyond Measure: A Spirituality of Nonviolence

The examples of the three markings on the table went from a mistake to mischief to mayhem with increasing intent to cause harm that corresponded to increasing damage. In the real world, intent and damage don't always correspond so neatly. A simple mistake can result in serious damage, injury, or death, and an attempt to create mayhem can result in a botched armed robbery with no physical damage done and no people hurt. What appears to be an accident might be well-crafted mayhem.

The mistake of not replacing a badly worn tire resulted in a brief scare on the highway and a long walk home for one teenager; for another teenager, a worn tire was the cause of the crash that killed her.

Mischief can leave little damage and a few laughs in its wake. Forty years after the fact, the semi-repaired tic-tac-toe mark on the walnut dining room table in our home is a great story starter for our kids. The images and sounds of their grandmother screaming not-so-polite Italian phrases while chasing her son around and under the table after he carved the markings into the family heirloom gets more detailed and creative with every telling. The marking is now a marker for a time past.

Mischief can also leave inconsolable grief. Leaving a school dance with a carful of friends, a seventeen-year-old driver jerked the wheel in an attempt to separate another couple who were necking in the backseat. The vehicle crossed the median and hit a van head-on. The van driver was killed, and his wife and five-year-old child were seriously injured. One teen in the car was killed, and one is paralyzed from the waist down. The driver is overwhelmed by the magnitude of the loss.

Planned mayhem is probably the most difficult to comprehend, to confront, and to heal from, since it leaves in its wake such meaningless, preventable destruction. There was no *reason* for it to happen, only lots of empty excuses ("He made me angry" or "She broke up with me, so I had to take it out on someone" or "She wasn't our kind"). The excuse given by one of the young men who savagely beat and tied twenty-one-year-old Matthew Shepard to a fence, leaving him to die, was that Matthew had made a pass at him. An eleven-year-old and a thirteen-year-old gun down four classmates and a teacher. A day after the killings, one boy cannot remember what happened; the other crawls up into his grandmother's lap and sobs. A group of teens beat a young girl and leave her to drown in a stream. They show no remorse and blame one another for the killing.

When youngsters create mayhem intentionally, or through their mistakes and mischiefs at home or in the community, neither harsh punishments nor full pardon will heal the victims or the perpetrators of the mayhem. It is nonviolent engagement that is at the heart of true reconciliatory justice: the willingness to confront wrongdoing and reach out to the wrongdoer. It refuses to allow us to divide our world and our relationships into "us" and "them." It denies us the myopic vision that limits our insight. It reminds us of our connectedness with one another and can point the way out of an impasse that bitterness and hatred have created.

In *Prisons That Could Not Hold: Prison Notes 1964–Seneca 1984,* Barbara Deming, a civil-rights activist, speaks about how "nonviolence gives us two hands upon the oppressor—one hand taking from him what is not his due, the other slowly calming him as we do this." The one hand keeps the offender from causing more harm to self or others; the other calms down the offender, allows time for reflection, and invites reconciliation. As our two hands reach out, there is at once an attempt to bring about a balance and a tension created that keeps both parties actively engaged in the reconciliatory process, as we strive to heal the rift created. We are attempting to restore community.

When a mistake results in serious damage, our arm of compassion reaches the farthest, while the arm of mindfulness helps the child acknowledge what has happened, confront the feelings of sadness, guilt, and fear, and take responsibility for the action, as well as rise above what has happened and get on with her life.

When mischief results in mayhem, the two arms are extended equally. Compassion and mindfulness are both equally needed.

When a child commits intentional mayhem, the arm of mindfulness is the longest at the beginning of the reconciliatory process, while the arm of compassion is still there.

The end goal in all three instances is an embrace in which the kids take responsibility as is warranted, are willing to make restitution, resolve to keep the mischief and mayhem from happening again, and commit to becoming once again active participants in the community. In the embrace, we are ready and willing to have them as participating members of our community.

For the youngster whose mistake or mischief has resulted in serious damage or harm, we might be inclined to offer only the arm of compassion. This will deprive the child of any opportunity to heal from within. She is likely to beat herself up emotionally, psychologically, and perhaps physically for the harm she has caused and can't fix. The young girl who was responsible for the death of a young father and a classmate, as well as the paralysis of another classmate, needed compassion and an opportunity to make both symbolic restitution (financial) and personal restitution (service at a rehab center). The compassion extended to her by her family and other members of the community helped her get through the months before her trial. The judge provided the arm of mindfulness. A healing embrace came from the father of the teen killed in the accident.

In an interview with Bill Johnson, a *Rocky Mountain News* columnist, the dead boy's father spoke of moving from shock to sheer frustration and anger, railing against God for the loss of his son's life. For eight months, he simply wandered. "I got up every day and hoped I'd make it through the next. By two o'clock every day, I was exhausted. Grief does that." A while later, the father went to visit the teen who had been paralyzed. Outside the hospital room was the young girl who had been driving the car. The dead teen's father reached out and embraced the girl. "She is a young woman who made a terrible mistake whose life will be forever scarred by what she did. I decided then she had suffered enough, and that I wouldn't be a part of continuing her pain by not offering compassion when she needed it most. In the courtroom, I wanted her to know in that place she was loved."

She was offered both the arm of mindfulness and the arm of compassion. She could begin her own journey of healing. She could once again be a participating member of her community. Before the dead teen's father could reach out to her, he had to give himself the time he needed to move from shock to frustration and anger, be immersed in intense sorrow, depleted of energy, and then finally be ready to get on with his own life for himself and for his other children. It was only then that he could reach out in compassion to the girl whose mischief resulted in the death of his son.

For the youngster whose intentional acts of violence and mayhem have resulted in serious damage, irrevocable harm, or death, we might be inclined to offer neither the arm of mindfulness nor the arm of compassion but rather the fist of vengeance and retribution. After the two boys, eleven and thirteen years old, shot their classmates and teacher, there were demands that the laws be changed so that the juveniles would have to spend the rest of their lives in prison. Some wanted the death penalty imposed.

When families and communities are faced with such violence, there are often cries for swift revenge and retribution, stronger punishments and stiffer sentences; it is hoped that these solutions will somehow "fix" the problem. Effective solutions are not that simple. Reconciliation is not found in homes and societies that rely on threats and punishments as primary tools to deal with mischief and mayhem. When the main goal is to make children (or adults) "pay dearly" for what they have done and to serve as "examples" for others who might think of doing the same, hate and bitterness find rich soil in which to grow.

In her book *Hard Questions, Heart Answers,* the Reverend Bernice A. King, daughter of slain civil-rights activist Martin Luther King, Jr., explains the problem with such a retributive society:

> Those who thirst for revenge may experience the illusion of satisfaction, but it never lasts long in people of conscience because every act of violence leaves in its wake the seed of more violence. . . . Revenge and retribution can never produce genuine healing. They can only deprive survivors of the opportunity for forgiveness and reconciliation that is needed for the healing process.

William Ayers, in his book *A Kind and Just Parent,* responded to the call for more severe punishment for young offenders by saying:

They are kids, after all, and nothing can possibly change them into adults. . . . And I want to will the Court—and then all of us—to set the highest possible standard when determining judgments: "If this were my child . . ." Nothing in that standard frees kids of consequences, nothing in it predetermines outcomes. It does, however, set a tone that is at once caring and complex.

How young offenders are treated will influence what kind of people they will grow up to be and what kind of lives the rest of us will live. If we don't help them reconcile with the community, we could well condemn ourselves to a lifetime of fear, distrust, and mayhem.

The following stories tell about ways individuals, families, and communities can and have responded to young offenders. When an entire community is committed to reconciliatory justice, young offenders are invited to rise above their misdeeds and violent acts. The goal is to mend and restore rather than isolate and punish. The search is not for vengeance but for ways to heal people and heal relationships.

We must finally be reconciled with our foe, lest we both perish in the vicious circle of hatred.

—Reinhold Niebuhr

A Town Totem and a Football Team

We use the elders to deal with the source of the problems, not just the symptoms, and to correct the imbalances in the community by healing both the victims and the offenders. The focus is on healing and restoration, not the adversarial process and punishment.

—Ovide Mercredi and Mary Ellen Turpel, *In the Rapids*

In the frenzy of homecoming activities, football players from one community burned the totem pole in the town center of

the opposing team. Excuses, accusations, absolutions, and outrage flowed like sap out of newly tapped maple trees. Those who tried to excuse the behavior of the boys and absolve them of any blame declared that they were "good kids, didn't understand the severity of their actions, didn't mean to harm anyone, just having fun that got a bit out of hand, thought the totem was merely a symbol of the town's football team." Others accused the boys of racial prejudice and wanted to see them punished severely. Some outraged adults saw the incident as an opportunity to seek revenge for past wrongs.

Rather than a fine and community service (the usual "punishment" for "good kids who do bad deeds"), the elders in the community with a burned totem offered a more constructive option to the judge. The boys could pay for the replacement pole. The man who had carved the original totem was willing to carve another and allow the boys to do the tedious work of sanding smooth the carving, all the while talking about the sacred significance of the totem for the townspeople. The time spent sanding was also to be used to discuss responsible and resourceful alternatives to totem burning, and ways to confront the racial prejudice and cultural ignorance that was prevalent in both communities.

When the totem was carved, sanded, and painted, the community held a more constructive "homecoming" ceremony. Young people pounded their drums in echo and counterbeat to the rhythmic heartbeat of the grandfather drum calling people from both communities to join in the healing celebration. Using soot from the burned totem, the elders blackened the now-callused hands of the football players, humbly washed them clean, and invited these young offenders to help unveil the new totem.

The elders had accomplished what no monetary fine or community service could: reconciliatory justice. The young offenders made restitution, figured out how they could keep such an incident from happening again, and healed with the parties they had harmed. With compassion, the elders helped make whole what had been rent apart.

Singapore Incident

In criminal law, decisions would be rendered from the perspective of healing, compensation, and reconciliation as opposed to Euro-Canadian values of punishment, deterrence, and imprisonment.

—Ovide Mercredi and Mary Ellen Turpel, *In the Rapids*

A few years ago, an American teenager, along with several friends, vandalized dozens of cars in Singapore. The Singapore government wanted to, and indeed did, cane the teenager. But not before the situation became an international incident. U.S. government officials, learning-disability-association spokespersons, and the boy's parents registered their complaints. "He was learning-disabled, he was bored, he was easily led, and he was just doing what he did to please his peers. It wasn't all his fault." What interested me in the incident was that neither the Singapore government officials nor those protesting the caning gave mention of repairing the numerous cars that had been vandalized. The response of these adults to the vandalism is a classic example of the two extremes: punishment and rescuing.

Under the guise of discipline and justice, physical (and emotional) violence toward the teenager is legitimized and sanctioned. Stripping the skin off a person's bare back and buttocks by whipping with a supple cane is punishment.

Using excuses, blaming a handicap, accusing others for the vandalism is rescuing.

With both sides arguing over, around, and past the teenager, it became quite clear that neither side wanted a true resolution to the problem. One side sought to instill fear and impose pain in order to "make him pay for what he did," while the other sought to absolve the boy of any responsibility for his actions. Instead of becoming an active participant in the dialogue and solution, the teenager became a footnote, albeit a caned one, to the news story.

Implementing the process of reconciliatory justice could have given the teen an active role in his own reformation and enabled those working with him, as well as those he harmed, the opportunity to be a part of reconciliation.

Restitution (fix what you did): You've vandalized several dozen cars. You have a job to do. All of these cars need to be repaired, and you need to figure out how you're going to pay for the damage you caused.

Resolution (figure out how you're going to keep this from happening again): You were bored. You're not going to be bored for a long while. You'll be busy earning money to make restitution. But there will come a time when you will have time on your hands. What are you going to do with it that is more constructive than vandalizing cars—work in a soup kitchen, help out at a nursery, volunteer with Habitat for Humanity? Don't tell me that you will never vandalize cars again; tell me what you are going to do with your free time in the future.

You're learning-disabled. You can use that as an explanation; please don't use it as an excuse. Excuses take away the power you have in your own life to take responsibility for your deeds and misdeeds.

You were easily led. You have two options here: You can figure out how not to be so easily led, or you can hang around kids less likely to lead you down the path you've just traveled.

You were just doing it to please your peers, to fit in with the group. What can you do that will help you have the courage to stand up for a value and against an injustice even in the face of peer pressure?

Don't just say you're sorry; show the people you harmed that you are actively working toward restitution, a change of heart, and a change of behavior.

Reconciliation (heal with the party you have harmed): What can you do to serve those whose cars you've damaged? Understand that some of those people will not let you touch their cars, so you might need to sweep their walks, wash their windows, run errands for them. Do something beyond mere restitution. It is good for your soul and good for your community.

There is nothing as easy as denouncing. It doesn't take much to see that something is wrong, but it takes some eyesight to see what will put it right.

—Will Rogers

Murder and Transformation

In the book *Jewish Wisdom,* Rabbi Joseph Telushkin retells the story first told by George Herald in "My Favorite Assassin," published in *Harper's,* April 1943:

But while the act of murder cannot be undone or pardoned, even a murderer can try to make his subsequent life worthwhile in other ways. The most powerful story about repentance I know concerns Ernst Werner Techow, one of three German right-wing terrorists who assassinated Walter Rathenau, Germany's Jewish foreign minister in 1922. The killers' motivations were both political extremism and antisemitism. When the police caught the assassins, two committed suicide; Techow alone survived. Three days later, Mathilde Rathenau, the victim's mother, wrote to his mother:

"In grief unspeakable, I give you my hand—you of all women the most pitiable. Say to your son that, in the name and spirit of him he has murdered, I forgive, even as God may forgive, if before an earthly judge your son makes a full and frank confession of his guilt . . . and before a heavenly judge repents. Had he known my son, the noblest man earth bore, he would have rather turned the weapon on himself. May these words give peace to your soul. Mathilde Rathenau."

Techow was released from prison for good behavior after five years. In 1940, when France surrendered to Nazi Germany, he smuggled himself into Marseilles where he helped over seven hundred Jews escape to Spain with Moroccan permits. While some had money, many were penniless, and Techow arranged their escapes for nothing.

Shortly before his activities in Marseilles, Techow met a nephew of Rathenau, and confided that his repentance and transformation had been triggered by Mathilde Rathenau's letter: "Just as Frau Rathenau conquered herself when she wrote that letter of pardon, I have tried to master myself. I only wished I would get an opportunity to right the wrong I have done."

You've done something terrible perhaps, but you're also a human being with a mind and a heart and a soul, and you've got to find a way to live beyond the worst thing you have ever done.

—Frank Tobin, teacher in the juvenile-justice system in Chicago

Grieving Dad Spurns Hatred

The man who opts for revenge should dig two graves.

—Chinese proverb

After his daughter, Julie Marie, was killed in the Oklahoma City bombing, Bud Welch just wanted to see the killer or killers "fry." However, almost two years later, Stephanie Salter from the *San Francisco Examiner* reported that the same Bud Welch who wanted the convicted bomber, Timothy McVeigh, to pay with his own life, had traveled over two thousand miles to spend time with Timothy McVeigh's father, Bill, and sister, Jennifer. In a phone interview with Ms. Salter, Bud Welch spoke of his journey toward reconciliation. "When I got ready to leave, Jennifer hugged me and then she just took to sobbing. I put my hands on her cheeks and held her face and said, 'Honey, the three of us are in this together for the rest of our lives. We can make the most of it if we choose. I don't want your brother to die, and I will do what I can to help.'"

Those two thousand miles were a short jaunt compared to the long road his heart would travel from April 19, 1995, the day his daughter and 167 other people were killed in the bomb blast of the Alfred P. Murrah Federal Building. For several months, he was filled with rage, revenge, and hate, wanting those responsible to pay with their own lives. Then one cold January day, he started to think about how miserable he was. "I was smoking three packs of cigarettes a day and drinking too much, and I didn't like myself. I wanted to know: After they were tried and executed, how was that going to help me? . . . I finally realized, it ain't going to help me at all. It sure won't bring Julie Marie back. Revenge, hatred, and rage—that's why Julie Marie is dead today."

Seeing the pain on the face of Bill McVeigh at the trial, Bud Welch wanted to tell him face-to-face that he cared about what this father of the man who killed his daughter was going through. After their emotionally charged meeting, Julie Marie's father told the reporter, "I've somehow felt closer to God than I ever have since I met with Bill and Jennifer. It was the most satisfying thing I have done in my life. It brought me so much peace, I can't tell you—and I'd recommend it for anybody who's been in a similar situation."

There are many who don't agree with him and have actually distanced themselves from him; some simply disagree with him and believe in their heart of hearts that retribution is the only way to go; others seem fearful that his presence might diminish their thirst for revenge. As if that diminishing of thirst for revenge would in some way dishonor the dead. With that thirst for revenge diminished, all that would be left is a pain-wracked heart. For others, their hatred is now embedded in their hearts and permeates their whole lives. They are not in possession of their hatred; their hatred now possesses them.

What Bud Welch came to know on his journey of reconciliation is that letting go of the need for revenge does not reduce the horror of the deed, does not excuse it, tolerate it, cover it, or smother it. He has looked horror in the face, called it by name, let it shock and enrage him; but his anger is now without hatred. "People still think you can get closure or healing from revenge, but you can't. I've had a really deep hole in my heart since Julie was killed. A big chunk of me is just gone forever. But killing Tim McVeigh isn't going to grow the chunk back." He has taken the energy of his anger and used it to fight the death penalty, himself now believing that vengeance has been tried for too long and produced only more violence and revenge. (Timothy McVeigh declared the bombing of the federal building to be retaliation for the U.S. bombings in Iraq.)

At least twice a week Bud Welch visits the "Survivor Tree," the American elm tree that his daughter used to park her car under on hot days. The tree was the only living thing left standing after the bomb blast. In an interview for the June 16, 1997, issue of *Time* magazine, Mr. Welch told a reporter, "When I go there, sometimes I lean against the trunk, close my eyes, listen to the leaves and think about the way it used to be. Then I go down to the fence, and strangers will sometimes ask me questions: 'Where was the front door of the building?' or 'Where was the truck parked?' Then I tell them who I am, and they share their deepest thoughts with me. That's a very positive thing—to touch and see and talk and visit. And to continue to tell the story of who my Julie Marie was."

And to be reconciled? It's as though you wake up one day and for the first
time in a long time, you're not encased in ice.

— **Ismar Schorsch**

A Nobel Peace Prize, a Good Friday Agreement, and a School

I forgive, but I remember. I do not forget the pain, the loneliness, the ache,
the terrible injustice. But I do not remember it to inflict guilt or some future
retribution. Having forgiven, I am liberated. I need no longer be determined
by the past. I move into the future free to imagine new possibilities.

— **Father Martin Lawrence Jenco,** ***Bound to Forgive***

On Friday, October 16, 1998, Catholic pacifist John Hume and Protestant David Trimble, onetime hardliner and unionist, were jointly awarded the Nobel Peace Prize for their efforts to bring about peace and reconciliation in Northern Ireland. Both men took part in the creation of the Good Friday Peace Agreement signed in April and overwhelmingly approved in May by voters on both sides of the conflict. Both men acknowledge that the agreement is a small step in the reconciliation process and that much more compromise and healing are needed before there is a lasting peace. And just as in other countries wracked by terrorism and violent discord, there is a need for the process of reconciliatory justice if the cycle of vengeance is to be stopped. More than thirty-five hundred people were killed during the twenty-nine-year conflict, including twenty-nine as recently as August 16, 1998.

As part of the Northern Ireland Good Friday Peace Agreement's promise to release more than four hundred convicted members of the Irish Republican Army and pro-British paramilitary, Thomas McMahon, an IRA bombmaker sentenced to life imprisonment, was set free shortly after the signing of the agreement. It was McMahon who planted a bomb that destroyed a yacht in 1979, killing Lord Mountbatten, the great-uncle of England's Prince Charles, along with Mountbatten's fourteen-year-old grandson, Nicolas; eighty-three-year-old Lady Brabourne; and fifteen-year-old boatsman Paul Maxwell while they sailed near Mullaghmore, in western Ireland.

According to a news report written by Shawn Pogatchnik of the Associated Press, the release of the convicted bomber stirred outrage among many in Ireland and England. Outrage was noticeably missing from the statement given by John Maxwell, the father of the fifteen-year-old killed in the attack.

"Keeping him in prison, will, unfortunately not bring my son back. Peace is the imperative now and we must look forward, so that perhaps Paul's death and those of thousands of others from both sides of the political divide here will not have been entirely in vain. . . . McMahon should not be kept in prison for sake of revenge."

This was not the first time that John Maxwell had looked forward since the death of his son. No matter how willing he was to let go of a need for revenge, John Maxwell daily faced the fact that the death of his son was forever a reality. Instead of letting his anger and frustration turn to hatred toward Thomas McMahon, he decided to take that energy and use it to fight the beast of terrorism. Education was his weapon of choice. He helped found an elementary school in Enniskillen, Northern Ireland, a school in which Catholics and Protestants are taught together.

Forgiveness is not a verb, nor is it an act of the will. For John Maxwell and other peacemakers, forgiveness is the voice of the heart that speaks the presence of the soul. It is heart business—the mind will be busy enough working out ways to demonstrate the forgiveness: through deeds, actions, releasing debt, and making real the tangible expression of forgiveness.

Reconciliatory justice is a visible expression of forgiveness and the act of healing in a community. It is perhaps the one tool that can begin to cut through the chains of violence. It does not excuse the violence, does not deny the dignity and worth of the victim or the humanity of the oppressor. It does justice to the suffering without perpetuating the hatred. It is the triumph of mindfulness and compassion over vengeance and retribution.

We cannot live for ourselves. A thousand fibers connect us with our fellow men; and among those fibers, as sympathetic threads, our actions run as causes, and they come back to us as effects.

—Herman Melville

Chapter 7

Life Lessons

The spirit is stronger than anything that happens to it.

—**C. C. Scott**

The book took almost two years to write. The stories are true. The lessons are universal and bypass cultural differences, language barriers, and borders. Suffering and loss are inevitable elements of life.

Recent events have once again made real the seemingly incompatible expressions that are three parts of the whole of living: Life is unfair, life hurts, and life is good. Also shown in living color are the three passages each one of us circles through on our journey—the piercing grief of good-bye, the intense sorrow as we reorganize our lives, and the sadness that shares space with a quiet joy and a gentle peace as we recommit ourselves to life, tempered by the loss and wiser.

On April 20, 1999, I watched in horror as relatives, neighbors, and my children's friends fled Columbine High School in our hometown of Littleton, Colorado. I watched in greater horror as some did not come out of the building that day. In the end, eighteen teenagers were seriously injured, and twelve teenagers and one teacher were killed by two students who then killed themselves. Our town was changed forever by two angry, alienated, disenfranchised boys who used assault weapons and homemade bombs to lay siege not only to their school but to the heart of our community as well. Once more we saw that bigotry, hatred, fear, and fanaticism have a human face and can cause just as much, if not more, suffering, pain, and grief as do the inevitable losses in life. Our personal and community grief was compounded because the suffering was intentional and unnecessary.

On April 28, in Taber, Alberta, Canada, a fourteen-year-old boy gunned down two students in the hallway of the high school, killing one and seriously injuring the other. On May 20, in Conyers, Georgia, six students were shot by a fifteen-year-old classmate, who then fell to his knees, stuck a gun in his mouth, and surrendered in tears. Two more towns are shaken and changed forever.

We share an unwelcome legacy of incomprehensible horror and sorrow with other communities. In the spring of 1996, a lone assailant opened fire at a primary school in Dunblane, Scotland. Sixteen children, their teacher, and the gunman died. Six weeks later, on another April 28, in Port Arthur, a town in Australia's island state of Tasmania, another deranged gunman etched a trail of blood as he sprayed bullets into a crowd of tourists. Thirty-five men, women, and children died, and scores more were injured. A year after the massacre in Port Arthur, Walter Mikac, a man whose wife and daughters had been killed there, wrote, "What happened there is not something you could ever contemplate." The Dunblane fathers said exactly the same about their town: "It was the last place on earth it could ever happen." But it did happen in their communities, and they were not the last to know such pain.

Only time will tell if the life lessons in these tragedies will be taken to heart. Will we point fingers, place blame, fortress our schools, seek revenge, and hold hostage another generation to the manacles of bigotry, hatred, and intolerance? Or will we, as individuals and an entire community, do what is necessary to take the weapons out of the hearts, minds, and hands of our kids? Will what we learn help us to create more caring, more compassionate, less alienating, less violent communities where our children can know, that they are welcomed as responsible, resourceful, resilient, contributing community members—each and every one having worth simply because they are children? Will these same lessons learned be put to use in Kosovo, Ireland, Rwanda, and Tibet, in our small towns, suburbs, and large cities? In this time of great chaos and suffering, can each casualty be given a human face? Can we reach out to others with compassion and empathy, honoring our deep bonds and common humanity?

On May 3, 1999, tornadoes ripped through Oklahoma and Kansas, killing almost one hundred people and injuring hundreds more; in a matter of minutes, entire communities were wiped out. As families buried their dead, helped their wounded neighbors, and sifted through the rubble that once was their homes, the strength and resiliency of the human spirit and the positive power of community joined together to help them begin to rebuild their lives. It is that same spirit and community power that helped the ice-storm victims of Quebec, the firestorm victims of California, and the earthquake victims of Mexico start over.

Can this deep passion to alleviate another's pain and sorrow become a part of our everyday life? Can we reach out to our neighbors who are suffering their own personal tragedies and ask, "What are you going through?" "What do you need?" "What can I do?" Can we be there for them as they name their loss, honor their grief, confront their pain, and tell their story? When we respond with a generous spirit, wisdom, discernment, abundant kindness, and mercy, when we help alleviate the suffering of others and we offer them our compassion and empathy, we create caring communities and safe harbors for our children.

When we offer our children our time, our affection, and our

sense of optimism, we help them find a way through their own adversity, grief, and sorrow. They learn that they, too, can take an active part in determining what they will do with what life has handed them.

In the midst of winter, I finally saw that there was in me an invincible summer.

—Albert Camus

Index